Living with Tourism

T0256271

Tourism is now recognised as a major player in the interaction between the local and global, especially within economic and social processes. Accordingly, significant attention is being paid to the practices and experiences of local communities as they are visited by and serve tourists. *Living with Tourism*, however, redefines the 'community' that is of interest when considering the effects that tourism has on culture, by providing an ethnographic account of both the toured and touring community.

Living With Tourism is an in-depth analysis of the interactions between tourists, the local community and place in the 'moonlike' cave-land of Göreme, a World Heritage Site in Central Turkey. It demonstrates the implications that community ownership and participation in tourism have for the politics of representation and identity, and also for the nature of the tourist experience. It is shown how, together with host communities, tourists themselves are continuously negotiating their own identities and experiences in interaction with the people and places they meet.

Living with Tourism develops a dynamic notion of culture and tourism sustainability, and therefore provides new insights not only for scholars of tourism, but also for those in the areas of anthropology, geography and social studies who wish to gain a deeper understanding of this global phenomenon in the contemporary world.

Hazel Tucker has a Ph.D. in Social Anthropology from the University of Durham and is a Lecturer in the Department of Tourism at the University of Otago, New Zealand. Dr Tucker teaches courses on tourist behaviour, social and cultural change and tourism, heritage interpretation and qualitative research methods. As well as publishing in the field of tourism, she has published articles in area of social and oral history.

Contemporary Geographies of Leisure, Tourism and Mobility
Series editor: Professor Michael Hall
Associate Professor at the Centre of Tourism, University of Otago, New Zealand

The aim of this series is to explore and communicate the intersections and relationships between leisure, tourism and human mobility within the social sciences.

It will incorporate both traditional and new perspectives on leisure and tourism from contemporary geography (e.g. notions of identity, representation and culture), while also providing for perspectives from cognate areas such as anthropology, cultural studies, gastronomy and food studies, marketing, policy studies and political economy, regional and urban planning, and sociology, within the development of an integrated field of leisure and tourism studies.

Also, increasingly, tourism and leisure are regarded as steps in a continuum of human mobility. Inclusion of mobility in the series offers the prospect to examine the relationship between tourism and migration, the sojourner, educational travel, and second home and retirement travel phenomena.

The series comprises two strands:

Contemporary Geographies of Leisure, Tourism and Mobility aims to address the needs of students and academics, and the titles will be published in hardback and paperback. Titles include:

The Moralisation of Tourism
Jim Butcher

The Ethics of Tourism Development
Mick Smith & Rosaleen Duffy

Tourism in the Caribbean
Trends, Development and Prospects
Edited by David Timothy Duval

Qualitative Research in Tourism
Edited by Jenny Phillimore and Lisa Goodson

Routledge Contemporary Geographies of Leisure, Tourism and Mobility is a forum for innovative new research intended for research students and academics, and the titles will be available in hardback only. Titles include:

Living with Tourism
Negotiating Identities in a Turkish Village
Hazel Tucker

Living with Tourism
Negotiating identities in a Turkish village

Hazel Tucker

Routledge
Taylor & Francis Group

LONDON AND NEW YORK

First published 2003
by Routledge
2 Park Square, Milton Park, Abingdon, Oxfordshire OX14 4RN

Simultaneously published in the USA and Canada
by Routledge
711 Third Avenue, New York, NY 10017

First issued in paperback 2014

Routledge is an imprint of the Taylor & Francis Group, an informa business

Transferred to Digital Printing 2009

© 2003 Hazel Tucker

Typeset in Galliard by Taylor & Francis Books Ltd

British Library Cataloguing in Publication Data
A catalogue record for this book is available from the British Library

Library of Congress Cataloging in Publication Data
A catalog record for this book has been requested

ISBN 978-0-415-29856-8 (hbk)
ISBN 978-1-138-00867-0 (pbk)

Contents

Acknowledgements   vii

1   Introduction   1
Göreme: an introduction 5
The wider context of tourism development in Göreme 10
Studying Göreme: the fieldworker, the field and the work 15
Structure of the book 20

2   Imaging Cappadocia: the construction of a tourist place   23
Scripting Cappadocia and Göreme 24
The 'lunar landscape' 27
Christian Göreme 29
Göreme as a cultural landscape 34
Touring Cappadocia 36

3   The tourists: in search of serendipity   43
Tourists in Göreme 44
Some tourist portraits 45
Tourists, non-tourists and the ill-fated plot 51
Avoidance of the pre-planned package tour 56
Budgeting for possibility 59
Temporal opening to possibility 61
Interacting with the authentically social 62
The search for serendipity, not touristic surrender 66

4   Continuity and change: gender and production in Göreme   69
Göreme lives 70
Göremeli/people of Göreme 75
Gender segregation: the duality in Göreme life 76
Memories of a life past: social change in Göreme 85

5   A community in competition: the business of tourism in Göreme   91
    Getting into tourism 92
    Tourism business and employment today 99
    Volcanic eruptions: competition, fights and gossip 103
    Consequences and experiences of tourism business 114

6   Close encounters: interactions between hosts and guests   118
    Packaged interaction 119
    The authentically social in Göreme village 120
    The power in hospitality 122
    Hospitality in the tourism realm 125
    Restricting hospitality 128
    Close encounters of a more 'real' kind 134

7   Romantic developments: new and changing gender relations
    through tourism   137
    Fun and romance 138
    Traditional gender relations 141
    Long-term relationships 143
    Caught in the middle 149
    Developing business through romance 152
    Developing romance: changing village life 156

8   The continuation of Göreme as a 'tourist site': politics of place
    and identity   159
    The discourse of preservation 160
    Villagers' experiences of preservation discourse 163
    The inevitability of change 166
    Emergent culture: the Flintstones of the future 169
    Experiencing the hypo-reality of Göreme 172
    The continuation of Göreme as a tourist site? 176

9   Conclusion   182
    Writing tourists into destinations 182
    Living with tourism in Göreme: a postscript 187

    Bibliography   190
    Index   208

List of Illustrations

Figures

Figure 1.1 Göreme village tourist map.   7

Figure 1.2 Map of Turkey highlighting the main route travelled
              by backpacker tourists.   10

Figure 2.1 Tourist map of Cappadocia showing the main
              tourist sites.   37

Plates

Plate 1.1 Göreme village.   6

Plate 1.2 The centre of Göreme village.   8

Plate 2.1 Miniature 'fairy chimneys' are sold at souvenir stands
             nearby full-size ones.   29

Plate 2.2 Byzantine frescoes in a 1,000-year-old church.   30

Plate 2.3 The coach park by the Göreme Open-Air Museum.   31

Plate 2.4 A hotel built to resemble the 'fairy chimneys' of
             Cappadocia.   40

Plate 2.5 The kind of vernacular accommodation that backpackers
             prefer to stay in.   41

Plate 3.1 Exploring in the valleys around Cappadocia.   58

Plate 3.2 Boiling *pekmez* (grape syrup).   64

Plate 4.1 Example of Esin's photographs.   74

Plate 4.2 Making bread to last through the winter is a full day's
                work, shared among a group of women.   84

Plate 5.1 A new tour agency is built.   98

Plate 6.1 'Turkish hospitality': a tea-house owner entertains
                tourists by playing his *saz*.   126

Plate 8.1 The National Park plan in action.   161

Plate 8.2 Erosion of the fairy chimneys.   169

Plate 8.3 Flintstones imagery used in Göreme's tourism.   171

Acknowledgements

I owe special thanks to the people of Göreme for their friendship, openness and generosity during and since the conducting of my fieldwork. I am especially indebted to the Köse family, particularly Abbas and Senem, whose support and kind hospitality were invaluable to my fieldwork. Many thanks are also due to Arif Yasa for his help with transcription of interview tapes, and to Mehmet Bozlak, Kaili Kidner and Lars Eric More for their practical support. I am indebted to all of the Göreme villagers I interviewed, including Mustafa Mızrak, as well as to the employees of the Ministries of Culture and Tourism and of the Göreme Tourism Co-operative who helped me in my research. I would particularly like to thank Dawn Köse, Andus Emge, the Tekkaya family, Ali Karataş, Hüseyin Uludag, Osman Atak, Mehmet Daşdeler, Hacer and Hikmet, Havva, Zubeyde, Zahide and Top Deck, for sharing information and friendship with me during my stay. Of course, the many tourists who shared their time and conversation with me also contributed greatly to this work.

In Durham, I extend my warmest thanks to Dr Tamara Kohn for her continuous encouragement, support and enthusiasm as I produced this work in its doctorate form. A number of people also contributed their time to discuss research and ideas, including Michael Carrithers, Mary Thompson, Arnar Arnason, Rachel Baker, Will Buckingham, Chris Fuller and Stephan Jamieson.

This book is largely the result of research funded by a scholarship from the Anthropology Department at the University of Durham. I am very grateful for this and other financial assistance provided by the British Institute for Archaeology in Ankara travel fund, the Durham University Council Fund for Students Travelling Abroad, and a Royal Anthropology Institute 'Sutasoma Award'. I would also like to express my appreciation to Ann Michael and Andrew Mould at Routledge for their support for this work.

My thanks go to Fez Travel for the use of the map of Turkey in Chapter 1, and to Matt Hall for his assistance in preparing the maps. All of the photographs are my own.

In the latter stages of working on this book, I have greatly appreciated the company and support of my colleagues and friends in New Zealand. In particular, I would like to thank Michael Hall for his wine and wisdom, Donna Keen for her assistance, Stephen Boyd for supplying the biscuits, Kirsten and Brent Lovelock, Anna Carr, James Higham, David Duval, Mel Elliot, Frances Cadogan, Linda Gilbert and Rob Scaife. Relief was also provided by John and his variety of beers at the Inch Bar.

Finally, I thank my parents, Carole and Jim Tucker, for their support and encouragement at every stage.

1 Introduction

In the Turkish village of Göreme where houses are dug out of rock, a group of five local men were planning to open a new tourism business called 'Bedrock Travel Agency'. When asked why they wanted to give it this name, one of the men answered 'Why not, Göreme *is* Bedrock, isn't it?' Many other tourism businesses in the village have also adopted this theme, following the American comedy cartoon in which 'The Flintstones' live caveman-style in rock-cut houses; Göreme has a 'Flintstones Cave Bar', 'Flintstones Motel–Pansiyon', 'Rock Valley Pansiyon' and so on. These businesses, dotted around the village, encourage a view among tourists of Göreme as a kind of fantasy-land purposefully adapted to accommodate themselves and their supposed tastes for 'different' and 'fun' worlds. Yet, while for them Göreme appears as something of a theme-park, seemingly created commercially for tourist entertainment and recreation, it is, from another view, a centuries-old village whose populace has long been digging caves for habitation out of the tall rock pinnacles that cover the landscape and working the dry rocky soil into gardens from which to live. Since the development of tourism in the Cappadocia region and the designation in 1985 of the Göreme valley as a World Heritage Site, the Göreme villagers have been treading the fine line between dealing with the often harsh reality of their own lives and simultaneously colluding with their tourist visitors to create the necessary fantasy to accommodate the visitors' desires.

Over the past two decades tourism has seemingly saturated both the physical and social fabric of the place. In a broad sense, this book, based on observations made in Göreme between 1984 and 2001, is an account of that change. Rather than assuming an 'impact' stance, however, *Living with Tourism* is an ethnography of both a *touring* and a *toured* community, which is intended to move beyond the often normative approach to tourism and socio-cultural issues in order to understand the actual cultural implications of the 'touristification of societies' (Picard 1996: 8). Tourism, as it is played out in particular places, is not only a meeting of different sets of people and each of their desires, intentions and practices, but it is also, inevitably, the *new* cultural forms and choices that arise out of such meetings. These new tourism-related cultural forms and choices engage a certain 'aestheticization' process (Selwyn 2001) that itself leads to further choices and practices.

The aestheticization of Göreme's 'moon-like' landscape of giant rock cones and the historic cave-dwelling culture embedded in the landscape is manifested in its World Heritage Site and National Park status. This marking of Göreme as a 'cultural tourism' destination by the heritage and tourism 'industries' has provided new choices and opportunities for the people of Göreme, leading them to think about and act upon their place and themselves in new ways. This book is about how the variety of people involved, including the tourists and the variety of 'mediators' (Chambers 1997) as well as the Göreme villagers, deal with and negotiate the new cultural identities, practices and relationships that inevitably emerge through tourism.

These are particularly salient issues today, and issues that are increasingly discussed not only in tourism contexts, but also in relation to shifting communities, processes of globalisation and what has become known as the 'global–local nexus'. Indeed, a point has been reached where our view of culture, as well as the ways that culture can and should be studied (Ahmed and Shore 1995), is having to be reformed. This is not only because of an increased need for relevance and applicability in the social sciences (ibid.), but also because of the ever-increasing number of shifts and transformations of culture taking place in the current state of 'globality', which render it no longer possible, nor indeed useful, for culture to be viewed as bounded in space and time (Clifford 1997).

Yet discussions about the ways in which 'traditional' and 'authentic' cultures are 'impacted' upon by tourism, and the ensuing need for measures of 'cultural preservation', are still frequently conducted by a variety of mediators and commentators, as if they were unproblematic. Coming from academic researchers through government officials to certain sectors of the tourism industry, terms such as 'delicate' and 'fragile', used to describe environments and cultures, and insistence on the need to maintain 'cultural integrity' are rife in discussions about sustainability in cultural tourism.[1] As Nuryanti (1996) has pointed out, studies of cultural tourism tend to be characterised by a series of perceived contradictions between the power of tradition, which implies stability or continuity, and tourism, which involves change. It is these perceived contradictions that have led much of the discussion on these matters to go round in ever-decreasing circles, creating the apparently unresolvable paradox that tourism destroys the object of its desire. It is little wonder that some have argued that terms such as 'authenticity' and 'sustainability' in tourism are meaningless.

As I will go on to show in this book, however, it is precisely because of their ambiguity and negotiability that these terms are not meaningless. The term 'sustainability' itself is loaded with the many different layers and angles of social, economic and aesthetic interest that the different people involved bring to each point of negotiation.[2] In its wider context, the concept of sustainability can be seen to arise out of the inequitable processes of globalisation, under which certain economies, cultures and environments are seen as being exploited, destroyed or, at best, subsumed by the homogenising forces of the West

(Mowforth and Munt 1998). Mass tourism itself is often seen as one of these homogenising forces. Yet it is the particular aesthetic and experiential dimensions of tourism, and the tourist's supposed search for some sort of experiential difference, that lead the processes of globalisation and homogenisation to be of such concern. To state the problem more directly (and in relation to the case of Göreme), is the obvious presence of tourism, for example in the selling of the local cave-dwelling cultural identity through the idiom of the Flintstones cartoon, ultimately changing and even 'ruining' the place, so that tourists will soon no longer find enough *difference* there and thus be driven to move elsewhere?

Similar questions are frequently asked in relation to tourism places anywhere, and they often engage the notion of 'authenticity'. The issue of authenticity, to be discussed at various points in this book and particularly in Chapters 3, 6 and 8, has had a prominent position in social theoretical discussions about tourists and tourism.[3] The problem is, however, that many of the ideas about tourist quest and experience, and their relation to the places and peoples that tourists visit, are merely assumed. A common assumption is that if tourists find something other than that which they had expected, something a little less 'authentic' and more 'touristified', they will be dissatisfied and will no longer continue to be interested in that place or experience. This also carries the assumption that tourist motivations and expectations remain fixed, and that is why destinations, host communities and cultural identities must remain static in order to satisfy them. These assumptions remain largely at the level of generality and speculative theory, however, and while tourism-related studies have provided increasing amounts of case material concerning local communities as they are visited by and 'serve' tourists, very little work has been done to include an ethnographic analysis of tourists themselves within and in interaction with the places they visit (Bruner 1995; Selwyn 1994).

Moreover, the assumptions about tourists' inability to accept cultural change and tendency to cling to the notion of primordial tradition or authenticity are surely themselves a part of the hegemonic discourse that promotes the preservation of the 'traditional' for tourist experience. Viewed in this way, such assumptions would seem to go hand in hand with the discourses on impacts and sustainability that, in their repeated conjuring of notions of 'fragility' and loss of 'cultural integrity' as mentioned above, can be little more than attempts by some commentators to disguise their own grievance over the perceived loss of 'traditional culture' as the grievance of the host society they describe.[4]

The attempt in this study, therefore, is to move beyond speculative assumptions by developing an in-depth analysis of tourism and tourists as they interact in a particular place. Such an analysis should be useful not only to illustrate many of the salient issues regarding the change that inevitably ensues from meetings between 'insiders' and 'outsiders' in tourism places, but also, and perhaps even more importantly, to investigate the structural conditions under which that change might be acceptable to the 'hosts' and 'guests' concerned.[5]

Göreme is particularly illustrative of these points because the tourism businesses there are largely locally operated and owned, and can thus be contrasted with the many tourist destinations that have become dominated by the intervention and insertion of exogenous business interest and capital. A nearby town that Göreme might be compared to is Ürgüp. Although Ürgüp is situated only ten kilometres away from Göreme, it has been shown by Tosun to be an example of 'unsustainable tourism development' at the local level (Tosun 1998: 595), because of the rapid growth of a large-scale tourism infrastructure that has pushed many of the smaller-scale locally owned operations out of business. The township, or village, of Göreme, in contrast, is relatively 'protected'[6] from large-scale capital investment and construction because of its proximity to the UNESCO-designated World Heritage Site of the 'Göreme Open-Air Museum' and its location within the boundary of the Göreme National Park. The Ministry of Culture imposes strict regulations concerning the construction of new buildings within the national park area and the building of large hotels is not permitted. In contrast to the situation in some other towns in the Cappadocia region, therefore, Göreme's tourism has remained relatively low on capital investment, especially investment coming from outside the village, and has developed in a pattern of small businesses that are mostly locally owned.

So, although this pattern of development is not the result of purposeful planning and design in that direction, Göreme might well be regarded as an example of 'community-based' tourism. The analysis of tourism in Göreme in this book is concerned with the implications of local participation and control in tourism development at the cultural level and in relation to tourist experience. Emphasis is placed on the role that community control in tourism plays in what might be termed 'the politics of representation and identity', and on the new tourism-oriented culture that inevitably emerges from the meeting between the local community and the tourists they play host to.

As a meeting between particular kinds of 'strangerhood', tourism always presents 'the problem of management of novelty and unfamiliarity in order to create order out of it' (Wang 2000: 148). For tourists, perhaps a part of the pleasure is in achieving successful management of this self-induced problem (ibid.). For 'locals', on the other hand, this management comes more as a matter of everyday necessity. Hence the widely discussed power imbalance inherent in the 'tourist gaze' (Urry 1990). In *Living with Tourism*, however, I show how the effective management of strangerhood by locals is possible when a performance and real working of hospitality can be, and is, brought into play. The analysis of tourism in Göreme thus challenges the paradoxical near-truism that cultural tourism destroys the very culture upon which it is based by showing how, when certain conditions prevail, it is possible for local communities to negotiate the continued success of tourism in the face of inevitable culture change and 'touristification' (Picard 1996). They can do so by asserting their position as 'hosts' and thereby negotiating their own and the tourists' identities in relation to each other.

Through eighteen months of anthropological fieldwork in Göreme, I have attempted to grasp the emic, or insider, perspective on the ways in which the

Göreme people *and* the tourists who visit Göreme understand and interact with each other. Recognising also that this interaction is very much mediated by social, cultural and economic processes that lie outside the immediate host–guest relationship, I look at the cultural 'baggage' that each of them brings to the meeting with the other, and at how that baggage determines the way in which they experience and deal with what each other is doing and what is happening in the place. *Living with Tourism* thus redefines the 'community' in tourism studies by investigating the multiple, parallel communities that meet through tourism, and by examining the interactions between them.

Göreme: an introduction

Two hundred kilometres south-east of Ankara in the centre of Turkey, Göreme is situated in the middle of a triangle formed by the three towns of Nevşehir, Ürgüp and Avanos, and lies at the meeting point of four valleys in the middle of the Cappadocia region. Named the province of Nevşehir in modern Turkey, Cappadocia was the ancient name for this region where the land comprises the out-spill of two volcanoes. The volcanic ash hardened to become *tufa*, a soft porous rock that has eroded over millions of years to form natural cones and columns, locally termed *peribacaslar* or 'fairy chimneys', on the landscape. For centuries the soft rock has been carved and hollowed to form the cave-dwellings, stables and places of worship that pattern the troglodyte village of Göreme today.[7] The climate in Cappadocia is hot and dry in summer (ranging between 30–40°C) and extremely cold in winter (reaching as low as −35°C). With frequent rains in the winter and spring, however, together with fertile soil conditions, the region has long subsisted on a traditional agricultural economy, with the most abundant and marketable crops being grapes and apricots grown in gardens and small fields.

Göreme has approximately 2,000 inhabitants, and can therefore register as a municipality and have a *belediye*, or municipality office, headed by a mayor. Since the downfall of the Ottoman Empire and the founding of the Turkish Republic in 1923, the village (previously known as Maçan but changing to the Turkish name of Avcilar in the 1950s, and finally to Göreme in the 1970s) has been inhabited by republican Turks, all of whom are Sunni Muslims. Throughout the first decades of the Republic, while the economy and infrastructure of Turkey underwent rapid adjustment and improvement, the economies of villages such as Göreme, situated on the Central Anatolian plains, remained near to agricultural subsistence. The hardship of this subsistence led to some villagers turning to the increasing number of outside possibilities of making a living and, in the 1960s, many of the area's inhabitants left to find work in northern European countries such as Germany, Holland and Belgium. For some Göreme villagers, trucking businesses also supplemented the agricultural economy throughout the 1970s and 1980s. Since the mid-1980s, tourism has developed to be the major source of income, although most families still keep up their farming practices.

Plate 1.1 Göreme village

The older residential quarters of the village are situated up the slopes away from the central village. The streets are steep, often narrow, and have a haphazard appearance with most houses built half into rock faces and 'fairy chimneys'. These cave-houses date back to the seventeenth and eighteenth centuries, but most were extended in the nineteenth century with Ottoman-style arched-room architecture constructed from cut stone added on to the original cave-dwelling (see Emge 1990; 1992). In certain areas, the older 'fairy chimneys' and cave-houses have been evacuated because of crumbling and rock collapse, and so those areas have a somewhat ghost-town appearance. Many of the families who left their crumbling older houses were re-housed in government-funded housing (*afet evleri* or 'disaster houses') that was built in the 1960s and 1970s in the lower end of the village. Since that time the village has continued to expand, with the ongoing construction of new concrete and brick housing in that lower flatter area beside the road to Avanos. Many of the villagers now living in that *Yeni Mahallesi* (New Quarter), however, still hold on to their cave-house in the older part of the village, and use the parts of it that remain intact to keep a donkey or a couple of cows, and also for purposes of food preparation and fruit-drying and storage.

Many of the older cave-houses also have been or are being restored as *pansiyons*, the local term for a guest-house, thus providing accommodation for tourists that is dotted throughout the older quarters of the village and is vernacular in style. Besides these, all of the tourism businesses, such as the restaurants, shops and tour agencies, are in the central area of the village and housed in

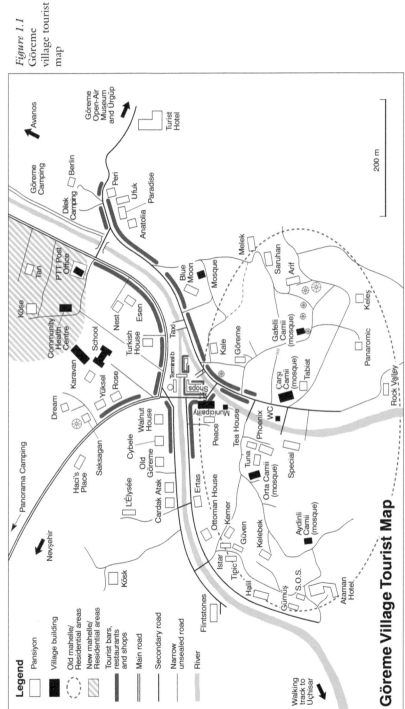

Figure 1.1
Göreme village tourist map

Plate 1.2 The centre of Göreme village: most of the tourism businesses are set around the bus station and shopping centre in this central area

newer buildings that line the central roads, acting rather like walls separating this central tourist area from the older residential areas behind (see Figure 1.1). Tourism thus presents something of an architectural and social duality in Göreme, and the distinction between the two areas or realms in the village is present in the minds of villagers and tourists alike, as well as the planning authorities who have a significant amount of control over building development and restrictions (see Chapter 8 for further discussion on this).

These planning authorities, both in the municipality office and the Ministry of Culture's 'Cappadocia Preservations Office' in nearby Nevşehir, have divided the map of Göreme into zones, with a clear distinction between a 'tourism' zone and a 'preservation' zone. The tourism zone, in the village centre, is where tourists are serviced with fun and entertainment, and where new building and tourism business is permitted and plentiful. The 'preservation' zone is up the winding residential streets away from the village centre. There, tourists and tourism are not so visible and the village women get on with their daily routine of domestic activities. Socially, this duality is prevalent in Göreme largely because of what Stirling earlier described in reference to other Central Anatolian villages:

> [a] strict segregation of the sexes and the fierce attitudes to feminine honour, which render it impossible for men and women to meet and co-operate except in and through their own households, or those of very close kin.
>
> (Stirling 1965: 98)

The women's domain is centred in and around the household, and so except for the time spent working in the fields or gardens in the valleys surrounding the village, women spend almost all of their time in their own neighbourhood (*mahalle*). It is inappropriate, as explained in depth in Chapter 4, for women indigenous to Göreme to work in the tourism business or to be present in the business sphere, and this in Göreme includes the *pansiyons*, the restaurants and the central area of the village generally. Consequently, there is little direct contact between tourists and village women, other than attempts made by a few women to sell headscarves or lace items to those tourists wandering by their gates in the residential streets. Tourism business, and the central area of the village, is thus almost entirely the domain of men. While they do occasionally partake in their households' agricultural work, the men mostly spend their days dealing with their business or, particularly in the winter months, sitting in the village tea house.

Around the tea house and lining the main road that cuts through Göreme, one can find the *belediye* (municipality) office, the central mosque (among a total of six mosques in the village), the school, post office, health clinic, tea house and a handful of grocery shops, including the village 'oven' (bakery) which produces endless quantities of bread and the smells to go with it. During the 1980s and 1990s this central area also became filled with tourism businesses such as *pansiyons*, restaurants, tour agencies, and carpet and souvenir shops.

When I first visited Göreme in 1984 there were three cave-house *pansiyons* and one small and basic restaurant in the village. Tourism had barely started to develop in Turkey then, and in Göreme it was almost non-existent. When I returned to Göreme in 1989, the village had over fifty *pansiyons*, as well as a handful each of restaurants, carpet shops and tour agencies. In 1990 a bus station, taxi rank and shopping complex were constructed in the village centre. Today, with a population of around 2,000 permanent residents, Göreme has approximately sixty *pansiyons* and motels, plus a handful of more upmarket hotels and a few camping sites. There is an 'Accommodation Association' office in the bus station where tourists can choose a place to stay on arrival in the village. Other tourism-related businesses include: fifteen or so tour agencies; fifteen restaurants; five or six bar/discos; fifteen carpet shops; several general stores; and numerous other souvenir shops and souvenir stands (these stands are mostly situated at prominent tourist spots outside the village, such as at the Göreme Open-Air Museum and a 'panoramic view-point' on the approach into the village from Nevşehir). There is also a horse ranch, which runs horse-riding tours, and a hot-air ballooning operation run by two pilots from northern Europe.

The main tourist attraction in the region is the Göreme Open-Air Museum. Situated two kilometres from the Göreme village, or township, of today, the museum is a narrow section of the Göreme valley where the particularly well-preserved remains of a caved monastic settlement dating back to the early Byzantine period are concentrated. In the late 1990s the museum received approximately half a million visitors yearly, three-quarters of whom were international tourists. The majority of these visit Cappadocia on 'cultural' package tours and stay in the large hotels in the nearby towns of Nevşehir, Ürgüp and Avanos situated outside of the Göreme National Park area. While coaches loaded with package tour groups drive through the centre of the village daily to reach the museum, the majority of tourists staying within Göreme village itself are young, lower-income tourists, travelling independently of package tours.

During the high season of the summer, the central area of Göreme village buzzes with tourist activity: tourists wander the streets and fill the many restaurants and shops that are there to service them; the little bus station in Göreme centre is crowded as hordes of tourists arrive off overnight buses in the early morning and leave by the same mode in the evening, headed mostly for either Istanbul or the south coast, depending on which way around they are 'doing the Turkey circuit'. Somewhere between 1,500 and 2,000 international 'backpacker' tourists, equal to the number of local residents living in the village, stay in Göreme every week during the busiest summer season, and there is a significant presence of tourists throughout the winter months as well.[8]

The wider context of tourism development in Göreme

Situated in the centre of Turkey, Göreme can be seen as being on the fringes of, and thus as feeding both off and into, the tourism boom in the Mediterranean. As a region, the Mediterranean is one of the most popular international tourist destinations in the world (Butler and Stiakaki 2001), but it is also riddled with

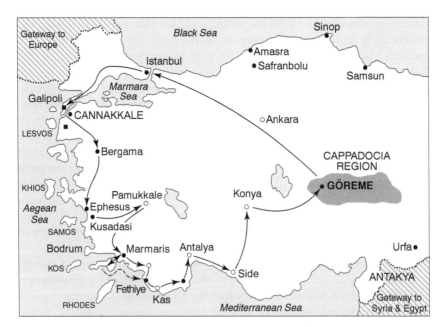

Figure 1.2 Map of Turkey highlighting the main route travelled by backpacker tourists

examples of what might be considered unsustainable tourism practices and destinations (ibid.). The rapid and large-scale tourism development in the region, based on sun, sea and sand, low prices and chartered flights, has led to a state of discontent that has driven many of the region's countries to diversify and promote other forms of tourism, such as those based on historical and cultural heritage, education and sporting activities. Turkey is no exception, and Göreme and the wider area known as Cappadocia have for the past two decades been a major focus of Turkey's 'cultural tourism' development.

The changes that Turkey itself has seen since its formation, led by Mustafa Kemal Ataturk in 1923, laid the crucial foundations for tourism to be developed. It has been noted by a large number of commentators that throughout the twentieth century Turkey achieved tremendous social and institutional reform and economic growth, with rapid industrialisation, infrastructure development and socio-economic change occurring throughout the nation. During the first years of the Turkish republic, the *per capita* income was US$47 (Öngör 1999). Any exports at that time were agricultural products, such as nuts, dried fruits and tobacco, and all industrial goods were imported. Seventy-five years on, industry and services make up a large share of the country's production, and the *per capita* income has risen to US$3,200 in the most conservative count (ibid.).

While this rapid economic growth led to mass urbanisation in Turkey, particularly in the latter half of the twentieth century, all outlying regions were also brought under the umbrella guardianship of the nation-state with effective reform and infrastructure development in even the most remote of Anatolian villages. Mustafa Kemal's republican authorities held the firm belief that a political revolution and the creation of the new nation was not possible without a social and economic revolution (Ahmad 1993). So, through massive state investment in roads, telecommunications, schools, electricity and health, the Kemalists managed literally to *build* the nation and to instil a powerful sense of nationalism among the large rural population (Bellér-Hann and Hann 2001; Shankland 1999). In the time-span of a few decades, the Kemalist reforms succeeded in undoing the Ottoman era's mindset, based on Islamic law, tradition and a clinging to the past, and replacing it with a society based on secularisation, modernisation and looking to the future.

In short, mass tourism development in Turkey stems entirely from the success of the political and socio-economic revolution that took place in modern Turkey throughout the twentieth century. Although the Turkish government had begun to look at tourism as an economic development strategy during the 1960s and 1970s, the rapid growth in tourism since then has undoubtedly been encouraged by the country's most recent economic development phase, which has been marked by decentralisation, liberalisation and an increased opening up to international trade (Fiertz 1996; Öniş 1996). It is also largely due to the relative political stability since the 1980 military coup, as well as to the positive image-building efforts by the Turkish government and the improvement of a tourism infrastructure, that tourism has increased so rapidly over the past two decades. For Turkey as a whole, foreign visitor arrivals increased from half a million in 1960 to almost two million in 1981 and then to over seven million in 1992 (Korzay 1994). By 2000 well over ten million international tourists visited Turkey annually, earning the country US$7.6 billion in tourism receipts (Travel and Tourism Intelligence 2001).

Following the 1980 coup, the new government under Turgut Ozal looked to tourism both as a tool for economic growth and as a means to creating a more positive and outward-looking image for the country in the international political arena (Sezer and Harrison 1994; Tosun 1998). Caught very much between East and West, Turkey is frequently depicted in academic and political discourse as a country torn between the two and not fitting comfortably with either (Nation 1996; Pope and Pope 1997; Spencer 1993). While the Turkish state looks largely towards Europe and the USA as its models and sources of economic development, a Western discourse of the dark and corrupt oriental Turk remains in place (Stiles 1991; Stone 1998; Wheatcroft 1995). Into the modern day this European image of Turks as 'barbarians' is reiterated in international publicity concerning human rights violations in Turkey, especially in relation to ethnic Kurds (Pope and Pope 1997: 38). Despite trade being rife between Turkey and Europe, and Turkey having an important strategic position for the USA and NATO, the West continues to keep Turkey at arm's length and Turkey is still some way from being allowed full entry to the European Union.

This Western image of Turkey and Turks as 'other', and the European association of Turkey with the exotic 'Orient' might, however, be said to do Turkey some favours regarding tourism. In addition to having a favourable climate and coast for beach holidays, Turkey is represented in travel literature as being rich in historical and cultural value. So, while much of the earlier tourism development took place around the south and west coasts, certain inland regions and towns were also identified under the Tourism Encouragement Act in the early 1980s as potential tourism centres. It was then that Cappadocia was identified as a 'cultural tourism' centre, and the Tourism Encouragement Act had significant implications for the way that tourism would develop there. This important piece of legislation ensured generous incentives for private tourism investment, while also annulling the prohibition of foreign companies acquiring real estate. As a consequence, large-scale tourism facilities grew rapidly in the region, particularly in the towns of Ürgüp, Avanos and Nevşehir.

According to Tosun, however, this growth took place largely 'in the absence of proper planning and development principles' (Tosun 1998: 595). As mentioned above, Tosun based his observations on the small town of Ürgüp, situated nine kilometres from Göreme. Ürgüp is similar to Göreme in its 'cave' setting and, following publicity created by northern European travel writers in the 1950s and 1960s, came to be visited by increasing numbers of independent tourists. Throughout the 1960s and 1970s, many local people opened *pansiyons*, restaurants and souvenir shops, and a 'friendly relationship' developed between the Ürgüp 'hosts' and their tourist 'guests' (ibid.). The new legislation enacted in the early 1980s, however, caused a dramatic turnaround in the situation of Ürgüp's tourism. Generous incentives to the large-scale sector of the tourism industry included guarantees of repatriation of foreign capital and no restrictions on the employment of foreign personnel. Foreign tour operators, together with national and international hotel chains, were quick to move in and, with their own marketing and the promotion of Ürgüp by the regional Ministry of Tourism and Culture office, Ürgüp became known as the 'tourism centre' of Cappadocia.

In this regard the two townships of Ürgüp and Göreme have long been in competition with each other, and this competition has particularly concerned claims over the patronage of the 'Göreme Open-Air Museum'. In its early days the museum was managed by the Ürgüp municipality, but later the site was appropriated by the regional government under the auspices of the Ministry of Tourism and Culture. It was then in the early 1980s, following the passing of a national law which stated that the municipality closest to any historical site could claim 40 per cent of the site's income, that the then-named Avcilar village appropriated the name of the museum and became Göreme. While the main reason for the name change was undoubtedly to have a stronger case in this claim on the museum's earnings, the change would also have served the purpose of attracting more tourists to the village itself. As one villager noted, 'It gave fame to this village, the museum and the name together.'

It was also because of its close proximity to the open-air museum site in the Göreme valley that Göreme village was included in the Göreme National Park area. The area officially became a national park in the mid-1980s, just at the time when the large-scale hotel developments were beginning to take place in Ürgüp. Situated within the park, Göreme village became subsumed under protection laws decreeing the preservation of all rock structures and houses, and severely restricting building and construction in the area. Unable, therefore, to obtain permission to build large hotels within or close to Göreme village, the foreign and national hotel chains built on sites outside the national park area, particularly in the nearby towns of Ürgüp, Avanos and Nevşehir. So, while Göreme remained relatively undisturbed by the 'mass' tourism moving into the region, these other towns saw the hasty construction of large three-, four- and five-star hotels. Today, most of the package tour groups visiting the region are accommodated in these larger hotels, and because the package tours are generally 'all-inclusive', many of the smaller, locally owned tourism-related businesses in Ürgüp and Avanos have been forced to close because of imperfect market competition (Tosun 1998).

The process by which small-scale 'informal' sector tourism businesses have been subsumed or ousted under pressure from larger 'formal' sector businesses has occurred in many tourism destinations around the world. Examples include Ladakh in India (Michaud 1991), Pattaya in Thailand (Wahnschafft 1982) and Kuta in Bali (Picard 1996). The similar chain of events occurring in Ürgüp has led Tosun (1998) to conclude that Ürgüp has undergone a process of 'unsustainable tourism development at the local level'. Indeed, in a conversation I had with a Tourism Ministry representative in the Tourism Office in Ürgüp about these matters, he said:

It's a debatable issue: Ürgüp townspeople do benefit indirectly from these hotels because the hotels buy food from the town and employ waiters and so on. On the other hand though, it is too quiet in town now, and the small restaurants and shops are suffering.

According to Tosun, this situation has occurred because all decisions related to tourism planning are made by central government, in a series of five-year national development plans, without consulting local governments. This planning follows a 'get rich quick' philosophy, and focuses on the economic benefits of tourism at the national level:

> This highly centralized planning approach to tourism development is the main source of problems in tourism development at the local level in Ürgüp, which, indeed, has planted the seeds of unsustainable tourism development.
>
> (Tosun 1998: 603)

It is interesting to note, in this context, that Göreme village remains almost unrecognised as a 'tourism centre' within official, popular and even academic circles at the national and regional levels. Göreme is a hive of activity in the tourism season, with the presence of tourists highly visible in and around the village, and with most businesses receiving a good amount of custom if not thriving. Yet I have been dismayed a number of times by tourism academics working at universities in Istanbul and Ankara and also tourism officials working in the regional office in Nevşehir asking me: 'Why are you researching tourism in Göreme? There is no tourism in Göreme.' One of the reasons for Göreme's low tourism profile is that, as I said above, Ürgüp has always been one step ahead of Göreme in its being viewed and promoted as the tourism centre of Cappadocia. Another reason is connected with the registration process concerning tourist accommodation. While hotels of more than ten rooms must be licensed nationally under the Ministry of Tourism and receive a star-rating according to their facilities, *pansiyons* or establishments with fewer than ten rooms need only register with and obtain a licence from the local municipality. *Pansiyons* are thus not fully recognised by the Ministry of Tourism and are not counted in national and regional tourism statistics. The majority of Göreme's tourist accommodation, therefore, is not visible within those statistics on which the level or amount of tourism and tourists in a particular place is usually judged in Turkey.

So, while large-scale business is recognised officially as being 'tourism', what may be regarded more as community-based tourism goes largely unnoticed within the broader political economy. Yet as we shall go on to see in the following chapters, tourism in Göreme, again in contrast to Ürgüp, might very well be regarded as a successful case of community participation in tourism, since the tourism business has stayed largely in the hands of the local villagers and those villagers consequently have a high level of control over tourist behaviour and representation regarding themselves and their village.

Studying Göreme: the fieldworker, the field and the work

As already suggested, the particular character of tourism development in Göreme, the complex set of relations occurring within it and the implications of this particular pattern of tourism development for the village, the villagers and the tourists make for fruitful case-study material that bears relevance to issues much broader than the Göreme situation itself. Of course, it should not be overlooked that ethnographic knowledge is always contextually tied to the location of the study and the methods used to collect it.[9] An ethnographic approach holds an advantage, however, in its ability to obtain a close-up and 'grassroots' perspective of the ways in which national and global trends and processes are experienced, negotiated and mediated by people 'on the ground' at the local level.

Accounts of the integration of Turkish village society into national and international political and socio-economic processes have been provided by a number of anthropologists and sociologists. One of the most thorough and useful works is Stirling's monograph *Turkish Village* (1965), which describes life in two villages in the Kayseri region of Central Anatolia in the late 1940s and early 1950s. Grounded in fieldwork, Stirling's work documented the processes of social and 'cognition' change as villagers become increasingly integrated into national society.[10] Bellér-Hann and Hann's recently published ethnography of the north-eastern part of Turkey named 'Lazistan' (Bellér-Hann and Hann 2001) also deals admirably with the effects that more widespread political and socio-economic changes at the national level have had at the regional, local and even personal level.[11]

Another ethnographic study of a Turkish town, and one that touches on the topic of tourism, is Mansur's *Bodrum: A Town in the Aegean* (1972). Although Bodrum is now a thriving (many would say overly so) seaside resort on Turkey's south-west coast, tourism had only just begun during the late 1960s when Mansur conducted her fieldwork. Her fieldwork period was also during the winter months and during her stay in Bodrum she lived in a hotel that was empty because it was outside the tourism season. She thus hardly makes mention of issues related to tourism, although in the postscript added to the monograph after revisiting Bodrum in 1971 she does comment on the changes that tourism was beginning to make to the town.

Most relevant to the present study is the work of Cemil Bezmen, whose doctoral thesis (1996) focused on the question of how Göreme's 'conservative' Islamic culture had managed to accommodate tourism. Bezmen's work was rigorous and dealt with the important issue of culture 'clash' through tourism, but it also left significant themes, and corresponding methods, still to be addressed. Being a male Turk and interested in the curious blend of religion and tourism, Bezmen largely focused his field research on the senior male members of the Göreme community. I, on the other hand, as a woman working alone, was able to spend time with the women of the village. I was thus able to gain a fuller picture of the gendered relationships and activities regarding tourism in the village. As a foreigner I also blended easily with, and was able to conduct participant observation among, the mostly international tourists. I was thus able to develop an in-depth understanding of tourist representations and experiences regarding Göreme.

My inclusion of the visiting tourists in my ethnography of tourism in Göreme is probably my most important point of departure not only from Bezmen's study of Göreme, but also from the other ethnographic studies that have dealt with Turkish villages or towns and social integration and change. By including an analysis of the tourists and their experiences, I have looked not just at social change *per se*, but at how the various representations of that social change come into play. As I suggested earlier, the extent to which tourism involves aesthetic valuing and experiencing should not be underestimated, especially where the negotiation of 'sustainable' tourism development is concerned. Even when those who first get involved in tourism business or policy-making are not fully aware of this point, they are soon made aware of it as their town or region competes with the other tourism destinations on the 'world stage'.

I had seen a huge boom in tourism growth in Göreme between the middle and the end of the 1980s, and when I returned to conduct fieldwork in 1995 I was struck by an acute awareness among villagers, tourists and other interested parties regarding how tourism, in all its aspects, was affecting the village. Through tourism, Göreme had not only become a meeting place of layered and conflicting representations and concerns, but also, necessarily, a focal point of vibrant discussion and negotiation. These were the issues, then, that my research questions turned to addressing during the fieldwork I conducted periodically throughout the latter half of the 1990s. I focused my attention on trying to understand not only what people were doing and how things worked, but also what people found important to talk about and the ways in which they talked, so as to try to understand how they experienced what was happening and how their various viewpoints interacted with each other.

My main stints of fieldwork in Göreme were two months in the summer of 1995, eight months from April to December in 1996 and seven months from March to October in 1997. They were followed by six weeks each in the summers of 1998 and 1999, and visits of just two and then one week in 2000 and 2001. I have thus been able to get a sense of the changes and continuities occurring over that time, to watch as new buildings were constructed and tourism businesses opened and closed, and to witness some of the events and politics taking place. During the main periods of my fieldwork, my Turkish language ability gradually improved so that I was able to converse adequately with villagers. Some village men working in tourism spoke with me in English, however, and this mix of languages is reflected in the interviews I carried out with villagers.[12]

Although I do not have room here to discuss at length the issue of reflexivity in ethnography, it is important to note that, as the anthropological 'self' is the 'research instrument' (Crick 1992), my own input into and experiences of the fieldwork should be understood alongside the ethnographic text and I therefore make no apology for where I include myself in this text.[13] For the most part, my fieldwork involved a combination of participant observation and interviews, both being contingent on the variety of relationships that I developed with different local parties and tourists in and around the village. Though I had already established some links with people in different sections of the Göreme community during previous visits, my introduction in 1996 to a village tourism entrepreneur named Abbas proceeded to have an important effect on my life and research in the village. I was offered the use and comfortable writing environment of Abbas's tour agency, situated on the main street of the village, in return for my help in answering the questions of enquiring tourists. I took up the offer for the first two months of my fieldwork, allowing me not only to chat with tourists coming in and to observe the workings of a tourism business in the village, but also to form a worthy relationship with Abbas and his family.[14]

Abbas had lived in Göreme all of his life and was one of the first men of the village to start up in the tourism business there. A middle-aged man and well-respected member of the community, Abbas adopted me as his 'niece' during my fieldwork and took it upon himself to take care of me while I was in the village. For most of my stay, Abbas arranged accommodation for me in a house that he rented for the tour guides who worked in his agency; he allowed me free access to his home, where I regularly ate, chatted and carried out domestic work with his wife, and he also allowed me continued use of the tour agency as a base from where I was able to observe tourist behaviour and tourism interactions.

A general understanding of ethnographic method is that fieldworkers gradually develop their understandings *in situ*, that the fieldworker ' "lives with and lives likes" those who are studied for a lengthy period of time' (Maanen 1995: 4–5). This method presents an interesting set of difficulties in situations where those who are studied are as diverse as local villagers and tourists on holiday. The finding of suitable 'observation posts' seems to be a crucial element in the ethnographic study of tourists and tourism, especially as tourists themselves are such a mobile and transient group. In discussing his fieldwork on tourism in Kandy, Sri Lanka, Crick (1992) talks about his relationship with 'Ali', a pavement hawker, and how the street corner where Ali sold his goods became an important observation post from where Crick could interact with and observe interactions with tourists. Similarly, Bowman (1996) sat with souvenir sellers in Jerusalem so as to observe their interactions with tourists.

The role of Abbas and his agency was especially important for my ability, as a woman, to live and conduct fieldwork alone in this Islamic village. As I discuss in Chapter 4, a crucial aspect of the gendered culture of the Turkish village concerns the Turkish verb *gezmek*, meaning to go out or tour around. While men may *gezmek* freely, it is improper for women to do so, and therefore women stay within strict boundaries of domestic space unless specific permission is granted by a close male relative to do otherwise. Due to the great many tourist women who travel through the region, the fact of my being a single woman alone there was not in itself an immediate anomaly to villagers. The problems arose gradually from my staying in the village for an extended period of time and from my research requirement to mix with people in both the tourism/public sphere and the domestic sphere of the village. The longer I stayed in Göreme, the more my identity transferred from 'tourist' to 'honorary villager'. And the more 'villager' I became, the more I was expected to conform to village rules, those of course being the rules for women.

Such conditions of fieldwork are discussed frequently by anthropologists.[15] However, my particular difficulty was one of how to carry out participant observation in the (male) 'tourism sphere' of Göreme, while retaining the respect of the villagers and the ability to continue establishing friendships with women. This is where the importance of Abbas and his agency came into play regarding their role in my reputation, and thereby my research, in the village. In villagers' minds the agency became the base to which I was attached, and Abbas the male 'relative' who gave me permission and protection whenever required. As I became known to be connected to Abbas, so he became the source of my ability to move around the village and to carry out my research in both the tourism and domestic spheres.

As my time in the village and my Turkish language ability progressed, various groups of women in various neighbourhoods gradually accepted me as a friend as well as a 'researcher' conducting a study into the life and customs of the village as they saw it. I often spent afternoons sitting chatting with women and girls in their houses, or, at times of harvest, helping in the fields and in their homes with the production of food. I also accompanied my women friends on some of their more formal visits to neighbours' houses, and to celebrations such as weddings, circumcision parties and the religious festival of *Kurban Bayramı* (Sacrifice Festival).

In the tourism sphere, I spent much of my time walking or cycling around 'visiting', drinking tea and chatting with the owners, managers and workers in their various tourism businesses. While sitting with village men in their businesses I could observe their behaviour and conversation as they interacted with tourists. I also spent much of my time with tourists in the evenings in the *pansiyons*, restaurants and bars, and during the days I would often join them on their explorations of the village back streets and their hikes through the Göreme valleys. I also periodically went on day tours of the Cappadocia region in order to observe that aspect of tourist experience in the area. While 'hanging out' with tourists in this way I was able to observe and note their reactions and interactions with villagers and the place in general.

My use and understanding of the Turkish language played an important role in my (participant) observations of interactions between villagers and tourists in that it enabled me to straddle insider/outsider identities. Tourists, for example, would see me as one of them, but sometimes also as a 'guide', or even confidante, with special 'insider' knowledge of Göreme society. On the other hand, I was able to be more on the 'in'-side at times when villagers were using Turkish to joke with, tease or secretly insult tourists. Of course, I was fully aware that my presence often affected the nature of tourists' experience and interaction. Despite such limitations, however, the methods of participant observation and informal interviewing allowed me to develop a much fuller picture of their interactions and experiences than the issuing of structured interviews or questionnaires could have done.

My undertaking of participant observation with tourists did mean, however, that the individuals observed tended to be arbitrarily rather than selectively chosen. Periodically I identified a particular range of nationalities, ages and 'types' of tourists for interviewing purposes, and developed specific sets of interview questions that enabled the development of certain themes which had arisen during periods of observation. Likewise, the research I carried out among local villagers followed a pattern of participant observation accompanied by the gradual development of interview schedules for either individual recorded interviews or focus group sessions with identified groups. I also carried out interviews with representatives of various authoritative bodies who have an involvement in Göreme's tourism, including agents of the national Ministries of Culture and Tourism, museum officials and the Göreme mayors.

My ability to meet with 'package group' tourists in the Cappadocia area was regrettably limited to a few 'chance' encounters and observations because of certain conditions associated with those tour groups. Bruner has also observed that it is 'difficult to penetrate the tour group from the outside at midpoint in their voyage' (Bruner 1996b: 162). The tour groups that visit Cappadocia and stay in the large hotels mostly situated on the periphery of the Göreme National Park are generally quite closed in the structure of their organisation. Although there may be a variety of reasons for this, in Cappadocia it is largely due to the fact that the souvenir retail outlets that the tour buses frequent, mostly selling carpets and rugs, silver and onyx to tourists, provide hefty commissions for each group's guide. The guides therefore try to avoid a situation where 'their tourists' might have contact with anyone outside the group for fear that they may be recommended to other, cheaper shops with which the guide has no commission arrangements. A commission system operates within Göreme village also – particularly among the carpet shops, but also the tour agencies – and this gave me some difficulties conducting my research because of expectations that I would lead tourists to particular friends' businesses. Although I did not accept commission, friendships I formed inevitably led to my allegiance with certain businesses and my unwitting position of competition against other businesses.

Indeed, the combinations and contradictions of 'self' and 'researcher' in the research context address the very core of the fieldwork experience and can often be difficult to work through. In Göreme, the often heated competition among tourism businesses, and the politics of tourism and village life in general, gave rise to many difficulties and dilemmas concerning both my fieldwork experiences and the writing-up of my research. The stakes run high and it is not surprising that many people whom I tried to interview were hesitant to publicise their views and experiences. Gossip, chastisement and even violence are rife in Göreme, and putting your foot in it can have serious consequences. In particular, my ability to be part of both the villagers' and the tourists' worlds, plus my position of straddling the quite separate worlds of the village men and tourism, on the one hand, and the village women and the domestic sphere, on the other, meant that answering my questions held potential risk for anyone.

Yet the time and help that I received in Göreme was given with seemingly limitless kindness and generosity. I have consequently experienced some of the feelings of anxiety that many anthropologists record regarding the taking of information from people with no significant return.[16] All I can hope for is that the people of Göreme, as well as the many others who participated in my research, are pleased with the product of the study in the form of this book, and with my attempts to shed light on some of their experiences there. Accepting, though, that it is primarily my fieldwork methods, relationships and experiences that are the locus of this research, I apologise for any sense of unfairness or misrepresentation that anyone may have. To the best of my ability, I have tried to present a fair account of tourism in Göreme based not only on my own understandings, but also on the understandings of the many people I had contact with during my fieldwork.

Structure of the book

Chapters 2 and 3 deal with the tourist perspective, focusing on tourist discourse, image and myth in relation to the Cappadocia region and Göreme. This organisation of the book is intended to take the reader immediately to an understanding of Göreme as a 'tourist place', thereby avoiding the presentation of a Göreme ethnography that the tourists appear later to invade or impact on. It also serves to emphasise the key idea that an understanding of tourists and tourist interactions should be seen as the base-point for discussions on culture and tourism sustainability in any particular place. In Chapter 2 an explanation of how the Cappadocia region became constructed as a tourist place shows not only how tourist representations of the region came into being, but also how the content of these representations became set in the tourist imagination and hence formed the tourist images and myths surrounding Cappadocia. An overview of the practices of package group tours of the region suggests that the primary tourist gaze, constructed by tour agencies and guides, is composed of these main images and myths. In addition, however, some representations of Cappadocia point tourists towards an alternative gaze, one that looks beyond the main tourist representations of the region.

This alternative gaze then forms the focus of Chapter 3, where the behaviour and experience of the independent tourists staying in Göreme is discussed. Through looking in particular at tourists' attempts to assert a 'non-tourist' identity, I suggest here that rather than being based primarily on a search for an experience of the 'authentic', the behaviour and experience of these independent tourists is based largely on a quest for serendipity. Hence, tourists' identities and behaviours are being continually negotiated in the context of their experiences and interactions in the places they visit.

The next two chapters switch the focus onto the people of Göreme and their varying levels and types of involvement in the tourism processes occurring in the village. In Chapter 4, I outline the socio-cultural processes underpinning tourism. The chapter initially focuses on gender relations in this predominantly Islamic village, and shows how tourism has encouraged the further social and spatial separation of a men's business sphere and a women's domestic sphere. The chapter then goes on to discuss socio-economic change in the village during the last five decades, looking at outward migration and the economic background to tourism development. Chapter 5 then discusses the development of tourism business in Göreme, examining the transition from a household agricultural-based society to one that is increasingly drawn into national and international economic tourism processes. It is shown here that although the particular development of small locally owned and managed businesses has promoted a culture of competition among the Göreme people, it has concurrently placed the villagers in a position of significant control over tourist interactions in their village.

The remaining chapters bring the tourists and the local population and place from the previous sections together to look at the interactions and outcomes of tourism. In Chapter 6, I explore the actual interactions that take place between tourists and Göreme men and women. While examining these interactions the discussion revisits concepts such as the 'tourist gaze' (Urry 1990), authenticity and hospitality. It is shown here that although the use of the terms 'host' and 'guest' in the tourism context has been criticised because it disguises the level of commercialism present in interactions with tourists, when the actual roles of 'host' and 'guest' can be played relations become altogether more complex and less one-sided. Chapter 7 then moves on to focus on longer-term tourists in Göreme and the sexual relationships between tourist women and local men. I also look here at the role of these relationships in tourism business development, as well as some of the consequences of this introduction of 'romance' for 'traditional' gender relations in the village. Chapter 8 then forms an analysis of the politics of representation surrounding the construction of Göreme as a tourist place, and identifies some of the conditions and parameters within which the negotiations of social identities and new cultural forms are played out. By drawing the main themes and issues discussed throughout the book together, the chapter considers the 'sustainability' of Göreme as a tourist site, thus bringing the book towards its conclusion.

In the concluding chapter, I summarise the analysis of tourism in Göreme and draw from it some more general conclusions relating to the wider theoretical themes raised throughout the book. I reiterate the need to re-think many of the prevailing views about culture and tourism, and also the value of undertaking in-depth ethnographic-style work on tourists as they interact with the peoples and places they visit. Only when we do so may we begin to understand the actual implications of tourism in societies and the ways in which the various groups of people concerned might be successful in dealing with those implications.

Notes

1 Pertinent examples of such discussions took place among the academics, politicians, and tourism industry and heritage organisation representatives who attended the International Conference on Heritage, Multicultural Attractions and Tourism at Bogaziçi University, Istanbul, in 1998 (Korzay, Burcoğlu, Yarcan and Unalan 1998). For further, more general examples, see Fagence (1998) and Zeppel (1998).

2 It was following the release of the Brundtland Report (WCED 1987) that the principle of 'sustainable development' really began to be embraced.

3 See, for example, Brown (1996), Bruner (1989), Cohen (1979, 1988b), Harkin (1995), MacCannell (1976, 1992), Moscardo and Pearce (1986), Redfoot (1984), Selwyn (1996), Silver (1993) and Wang (2000).

4 In other words, social scientific research on tourism has been characterised by the view that tourism causes the destruction of the 'genuine' socio-cultural elements of host societies. While some discussions about tourism and culture have now reached a more advanced stage of maturity (examples of which are Chambers 2000; Lanfant *et al.* 1995; Mansperger 1995; Picard 1996; Wood 1997), this earlier view continues to be expounded today in academic and, even more so, popular discourses.

5 Craik (1995) has established a need to consider 'the cultural limits of tourism'.

6 To use the terminology of the 'Cappadocia Preservations Office' in the regional town of Nevşehir. This organisation operates under the national Ministry of Culture, and is discussed further in Chapter 8 of this book.

7 Göreme is actually a *kasaba*, which is a Turkish term for a place between the size of a village (*köy*) and a town (*şehir*). With no suitable translation of *kasaba* in the English language, however, I refer throughout this book to Göreme as a village, since the people of Göreme themselves usually refer to the place as their village and to themselves as villagers (*köyluler*, which also translates as 'peasants').

8 These figures were calculated in 1998 using the sales figures of the long-distance bus companies in Göreme, which are the predominant means for independent tourists to travel to and from Göreme.

9 See, for example, Brewer (2000), Davies (1999), Ellen (1984), Maanen (1988, 1995) and Okely and Callaway (1992).

10 See Stirling (1965, 1974, 1993).

11 In addition to these major ethnographic works, other in-depth studies of Turkish society and social change have included Ari (1977), Engelbrektsson (1978), Kocturk (1992), Magnarella (1974), Stokes (1992), Tapper (1991) and the chapters making up Stirling's 1993 collection. Dealing more closely with the topic of gender in Turkish society are Delaney (1991), Kandiyoti (1991), Marcus (1992) and Scott's work on gender and tourism in Northern Cyprus (Scott 1995, 1997).

12 Direct quotes from interviewees presented in this book are marked '[*trans.*]' where they are translated from Turkish into English.

13 For discussion on reflexivity and representation in ethnography, see Bell (1993), Davies (1999), Hertz (1997), Maanen (1988, 1995), Okely and Callaway (1992) and Woolgar (1988).

14 With the exception of Abbas and Senem Köse, and Mustafa Mizrak, pseudonyms have been used throughout the book to ensure anonymity.

15 See, for example, Golde (1986) and, in relation to fieldwork in a Turkish village, Delaney (1993).

16 See, for example, Hastrup's comments on the 'violence' inherent in fieldwork (Hastrup 1992) and also Jackson's discussion based on interviews with anthropologists about their fieldwork experiences (Jackson 1995).

2 Imaging Cappadocia
The construction of a tourist place

Daylight, as usual, saw us well on our way. As we slowly rose to the top of Topuz Dag, a magnificent sight burst into view. Before us spread out a vast expanse dotted with multihued sugar-loaf cones, some the size of an ordinary tent, others the height of lesser skyscrapers. It is said fifty thousand cones can be counted, and, whether this estimate is correct or not, the figure does not overestimate the impression....The view was not only of magic form but also of vivid colour. The crags, cliffs and cones varied from snow white to cream, tan, ochre, pink, red, and grey. The very atmosphere seemed steeped in brilliant hues.

> (J.D. Whiting writing on his arrival in Cappadocia – Whiting 1939: 763)

Since the time that Whiting journeyed to Cappadocia the region has most definitely become marked out as a tourist place, so that now over half a million tourists visit yearly.[1] Entering the Göreme valley for the first time, visitors today also appear awe-struck by the 'moonscape' of valleys and fairy chimneys stretching out in front of them. Unlike Whiting, however, these contemporary visitors inevitably carry in their imaginations ideas and images of the Cappadocia region taken from guidebooks, magazines and television, tourism brochures and, increasingly today, from the Internet. What Whiting probably did not realise is the part that his writings, among other early travel articles, would play in the forming of these images that today inspire so many tourists to visit Cappadocia and that have culminated in the production of the region as a 'tourist place'.

This chapter is an introduction to how Cappadocia became, and continues to be, a tourist place. The discussion shows not only how tourist representations of the region came into being, but also how the content of these representations became set in the tourist imagination and hence formed the tourist images and myths (Selwyn 1996) surrounding Cappadocia. These images and myths make up the primary tourist gaze (Urry 1990) that the tourism and heritage 'industries' project onto the region, establishing what should be seen as extraordinary and thus worthy of having the tourist gaze cast upon it. I will go on to suggest at the end of this chapter, however, that this is not the only, singular tourist gaze on Cappadocia, and I will introduce an alternative gaze that I will then elaborate on in the next chapter. As MacCannell has recently noted, although the guidebooks and brochures denote for us what is worth viewing, 'we remain free to look the other way, or not to look at all' (MacCannell 2001: 24). Indeed, tourists often do look the other way, and I will go on later to discuss this 'other way'. First, though, I will consider the content of some of the tourist literature on Cappadocia in order to see what it does mark off as extraordinary and worth viewing.

Scripting Cappadocia and Göreme

Early representations of the Cappadocia region in travel articles, such as that by Whiting (1939) and an even earlier photographic essay in *National Geographic* (1919), served to 'script' the region for future tourist representations. Contemporary articles and guidebooks continue to follow the markers set out for them, using the same methods of description, and assigning the landscape, people and history the same mythological value as the earlier articles. As Gregory has pointed out, each successive visit and representation to a place:

> contributes to the layering and sedimentation of powerful imaginative geographies that shape (though they do not fully determine) the expectations and experiences of subsequent travellers.
>
> (Gregory 1999: 117)

Moreover, when such representations are used in tourism promotions they are intended to woo the tourist imagination by:

> drawing attention with an exotic element, and then – having captured their readers' attention – inviting them to imagine how they might feel in the setting depicted.
>
> (Lutz and Collins 1993: 3)

Tourism, in this way, becomes a repeated search for sights that other tourists have identified as unique, exotic or significant.

Earlier representations of Cappadocia have thus played a strong part in the construction and perpetuation of the images and myths existing today in the tourist imagination. Following are extracts from four different contexts in illustration of this. One is an American travel article from the 1950s, one an extract from a mid-1990s Turkish government tourism promotion magazine, another from a late 1990s British tourist brochure, and the last from a website written by a Turk promoting Cappadocia. Although these examples are, of course, by no means a comprehensive coverage of the variety of tourism-related literature on Cappadocia, they demonstrate how the content, written and pictorial, of the range of tourist literature forming representations of Cappadocia, whether past or contemporary, foreign or Turkish, tends to highlight the same features of the area. These features are the volcanic landscape of valleys filled with rock cones, the Christian (Byzantine) history, and the contemporary 'peasant' and cave-dwelling way of life.

The first caption accompanied a photograph essay entitled 'Cappadocia: Turkey's Country of Cones' in the *National Geographic* of 1958:

> Fantastic rock cones stud Göreme Valley in the heart of the Anatolian highland. Slumbering on the horizon, snow-capped Erciyas Dagi, Asia Minor's loftiest peak, guards a conical brood. This volcano, now extinct, created Cappadocia's coneland. Eons ago the fiery mountain spewed ash and lava hundreds of feet deep across the countryside. As the mass cooled, it cracked. Rains and melting snows widened seams into chasms and carved pyramids. Early Christians hollowed cells and chapels within the cones, and today Turkish farmers dwell in many of the honeycombed rocks.
>
> (*National Geographic* 1958: 123)

The second extract is from a 1995 Turkish Ministry of Tourism promotional magazine for the Ankara and Central Anatolian Region:

> Volcanic eruptions of the volcanoes Mt. Erciyes and Mt. Hasan three million years ago covered the plateau surrounding Nevşehir with tufa, a soft stone comprised of lava, ash and mud. The wind and rain have eroded this brittle rock and created a spectacular surrealistic landscape of rock cones, capped pinnacles and fretted ravines, in colours that range from warm reds and golds to cool greens and greys. Göreme National Park...is one of those rare regions in the world where the works of man blend unobtrusively into the natural surroundings. Dwellings have been hewn from the rocks as far back as 4000 BC. During Byzantine times, chapels and monasteries were hollowed out of the rock, their ochre-toned frescoes reflecting the hues of the surrounding landscape. Even today cave dwellings in rock cones and village houses of volcanic tufa merge harmoniously into the landscape.

The third extract is taken from the Tapestry Holidays 1997–8 brochure entitled *Uncommercial Turkey*. This brochure provides six pages on 'Cappadocia – a captivating playground':

> Subterranean cities, cave houses, tunnels, fairy chimneys and an enchanting landscape – what a playground!...Although renowned world-wide we find that many Britons are not aware of the huge historical and religious importance of the area and also its fascinating, strange 'moonlike' geology...The natural structure of Cappadocia offered refuge with the texture of the rock allowing the Christians to build cave houses, tunnels and even subterranean cities relatively easily...The history, culture and amazing scenery, together with the most hospitable people, proud of their heritage, make Cappadocia a region that will interest not only those of an historic or religious nature but is perfect for those of us that enjoy walking and relaxing in wondrous calm, for children who can explore the tunnels linking one valley to another

in absolute safety and generally all that would wish to visit a truly unusual corner of our planet.

The last example is quoted from a website on Cappadocia, written by Dağhan Erdoğdu:

> Besides its unequalled and striking scenery, Cappadocia is full of artistic products belonging to different civilisations. An active rural life, with all its authentic colour and folklore, completes the atmosphere. Most of the monasteries, churches and monk cells are decorated with frescoes. Nature, history, art and life itself meet nowhere else in such harmony and unity. Once this magnificent sight was likened to the 'surface of the moon'.
>
> (Erdoğdu 2002)

The photographs used as illustrations in all of these articles depict the same key features of Cappadocia that are emphasised in the text. All show panoramic views of the 'lunar' landscape as well as closer views of individual fairy chimneys. They also show samples of frescoes painted inside the caved Byzantine churches, and some photographs depict contemporary life inside the rock cones and caves, or scenes of rural 'peasants' on donkeys going to their fields. The selection and manipulation of these particular images form the myths that are the very foundation of tourism in Cappadocia. According to Selwyn (1996), tourist images are linked to myths in that myths simultaneously reveal or 'overcommunicate' some features of a place or people while concealing or 'undercommunicating' others. Certain features of Cappadocia, denoted as exotic and unique, are overcommunicated, and others, those considered to be mundane and ordinary, are omitted. Wang (2000), following Berger (1972), calls this the 'tourist way of seeing', suggesting that it is quite different from other ways of seeing, such as political, scientific or 'local' ways. Wang (2000: 161) argues that the characteristics specific to the tourist way of seeing are that it is apoliticized, decontextualising, simplifying, ahistoricizing and romanticizing.

The Romantic Movement, defined as the 'cultural reaction to and a critique of modern capitalist industrial civilisation' (Wang 2000: 86) that developed in the West during the eighteenth and nineteenth centuries, has had a considerable influence on tourism (Ousby 1990; Wang 2000). This influence is evident in the tourist images and myths surrounding Cappadocia, particularly when the different manifestations of Romanticism are considered. Wang (2000: 87) cites Mumford's classification of the romantic reaction into three groups (Mumford 1934): the cult of history, the cult of the primitive and the cult of nature. All three are related to tourism in that they lie behind tourist quests for particular experiences and the marking of certain sights, places and cultures as worthy of tourist viewing. The cult of history is born out of the nostalgic yearning of Romanticism and translates in tourism into the 'heritage industry'. The cult of the primitive is the search for people still living a simple, natural life associated with pre-modern times and gives rise to ethnic or cultural tourism. The

Romantic cult of nature is a yearning for 'natural nature', which 'acts as the "green" dream-place in contrast to the "grey" urban nightmare' (Wang 2000: 89). This yearning is echoed in rural and nature tourism and the construction of 'landscapes' and 'wilderness' for tourist consumption. All three of these romantic cults can be seen in the tourist images and myths concerning Cappadocia. In the next sections I will discuss each of these in turn, starting with representations of the natural landscape.

The 'lunar landscape'

The very idea of nature as landscape is socially and historically specific, and is linked, as suggested above, to the Romantic Movement and the 'cult of nature'.[2] Not only does the term denote the picturesque, but it also encompasses the extraordinary and the 'sublime', which represent those parts of nature that have not yet been fully controlled by humankind (Wang 2000). Photographic and written images of the physical landscape of Cappadocia, describing the landscape as 'fantastic', 'bizarre', 'unique', 'surreal', a 'moon-scape' and a 'fairyland' serve to attract tourists to view its 'otherness'. These representations are layered on one another, with earlier descriptions being incorporated into later ones as if somehow to authenticate not only the landscape itself but also the traveller's visit to and representations of the landscape:

> As you enter Cappadocia...it's obvious. Père Guillaume de Jerphanion, a French traveller who beheld the scene in 1905, wondered if he was 'looking at a real landscape' or had been 'transported, by some prodigy, to the most improbable fairyland scene. Emerging here and there from a green land mass is a pyramid, a cone, a tower that one would believe to have been made by the hand of man, so regular is it, and which, on examination, appears to be nothing but rock'. We were surrounded by surreal vistas of serrated geology...hundred-foot-high pinnacles prodded the sky, pinkening with the sunset...In the distance, the broad, purpling plateau of Anatolia – the ancient name for Turkey – was cut with valley after valley of cones.
>
> (Holmes 1997: 56)

Not only has the landscape of Cappadocia been marked off as worthy of tourist interest by these representations and images presented in past and present travel and tourism literature, but the representations in turn show tourists what to expect, what to look for and how to read the landscape. This can be seen in the descriptions of the Cappadocia scenery from earlier foreign travel articles through to more recent Turkish ones:

> Nature's 'sand blast', storm winds bearing coarse grit, gives the cones a polished appearance. Tints vary from snow white to cream, tan, pink and grey.
>
> (*National Geographic* 1919: x)

The view was not only of magic form but also of vivid colour. The crags, cliffs and cones varied from snow white to cream, tan, ochre, pink, red, and grey. The very atmosphere seemed steeped in brilliant hues.

(Whiting 1939: 763)

The wind and rain have eroded this brittle rock and created a spectacular surrealistic landscape of rock cones, capped pinnacles and fretted ravines, in colours that range from warm reds and golds to cool greens and greys.

(Turkish Ministry of Tourism 1995)

Evident in these extracts is the point made by Wang that 'the works or discourses of the romantics formed a perceptual framework or a structure of perception through which tourists came to see landscape' (Wang 2000: 85). The tourist perceptual framework is primarily an *aesthetic* one rather than a utilitarian one, and this aestheticization of landscape is the reason why the process of 'creating tourist places embeds the tension of commodities into the landscape' (Sack 1992: 157). First, implicit in the tourist way of seeing landscape is the suggestion that the landscape is there primarily for the purpose of tourist – aesthetic – viewing. Indeed, underpinning all of the representations of Cappadocia in travel and tourism literature is the idea of the tourist's prerogative to consume the landscape. The Tapestry Holidays *Uncommercial Turkey* brochure quoted earlier certainly conveys this idea: 'Subterranean cities, cave houses, tunnels, fairy chimneys and an enchanting landscape – what a playground!' The Turkish Ministry of Tourism promotional magazine states that 'The town of Göreme is set right in the middle of a valley of cones and fairy chimneys', as if it had been put there for the convenience of the visiting tourists. The construction of the landscape surrounding Göreme as a national park also conveys the idea of the area's purpose being for tourist consumption.

Second, being an aesthetic way of seeing the tourist gaze is constantly engaged in aesthetic scrutiny and judgement. Urry (1992) has pointed out that for landscapes to be suitable for tourist consumption, they must be unique, unpolluted and authentic, although he does also point out that these terms are negotiable. Representations of the Cappadocia landscape in the tourist literature certainly proclaim its uniqueness:

The peculiar formations and sights of the region are definitely unique. One cannot help feeling that some majestic sorcerer has chosen this place to perform his magical wonders.

(Erdoğdu 2002)

The 'other-worldliness' of the rock cones was even used for some scenes in the film *Star Wars*.

While uniqueness allows for some level of objective measure, the qualities of being 'unpolluted' and 'authentic' necessitate a particular kind of scrutiny and protection against what are considered to be polluting or de-authenticating

Plate 2.1 Miniature 'fairy chimneys' are sold at souvenir stands nearby full-size ones.

influences. The importance placed on the Cappadocia landscape in its continual representation in the tourist literature has manifested in the formation of the Göreme National Park and the Cappadocia Preservations Office in the nearby town of Nevşehir. These two organisations together pronounce strict regulations aimed at 'protecting' the landscape from polluting elements, such as tall buildings or anything and anyone that would damage the existing rock formations. Cappadocia has become a tourist place because of the aestheticization of its 'lunar landscape', and so that aestheticization needs to be continually pampered so as to be upheld. These issues are discussed further in Chapter 8.

Christian Göreme

According to much of the tourist literature on Cappadocia, the region was 'discovered' by the West in the early twentieth century when a French priest named Guillaume de Jerphanion conducted and published a study of rock-cut churches in the Göreme valley:

> His findings are contained in a massive work, *Une nouvelle province de l'art byzantin: Les églises rupestres de Cappadoce (1925–1942)*. Jerphanion's lead

was followed by other scholars and the importance of the churches was recognised. Today many have been designated archaeological monuments and are carefully maintained and preserved.

(*Blue Guide to Turkey* 1995: 612)

Jerphanion's work served to mark off the Byzantine churches in the Göreme valley as being of key historic significance. Other writings and photographic representations from the early twentieth century, such as the photographic essay in *National Geographic* of 1919, emphasise the historic and visual significance of both the churches and the frescoes on their rock-carved walls, thus denoting their value for tourist interest. Contemporary travel guidebooks and tourist brochures all repeat this emphasis with descriptions and photographs of the frescoes in the churches considered to be the most important. The *Blue Guide to Turkey* (1995), for example, fills twelve of its fifteen pages of the Cappadocia section, and the *AA Essential Explorer Guide to Turkey* (1995) five out of thirteen Cappadocia pages, with thorough description of the key Göreme churches.

Comparatively, the eras previous to and since the Byzantine period are mentioned only briefly in the tourist literature. According to historic interpretation numerous tribes, including Hittites, Persians, Syrians and Armenians, have

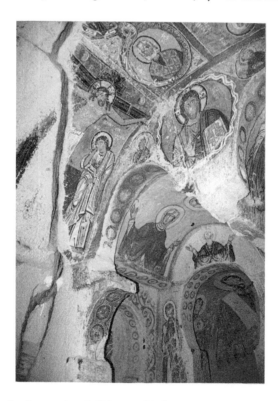

Plate 2.2 Byzantine frescoes in a 1,000-year-old church.

settled the steppe-lands of Central Anatolia. The Byzantine period in Cappadocia's history is still very much physically and visually present today, however, with approximately 300 cave churches and monasteries, dating from between the nineth and the thirteenth centuries, still remaining in the area. These churches are scattered throughout the valleys in the entire region, but the Göreme valley, studied by Jerphanion, is a particularly concentrated area of monastic settlement. It was this site that thus became marked off and enclosed as the Göreme Open-Air Museum in 1950. The area was designated a UNESCO World Heritage Site in 1985, and since that time UNESCO has part-funded the restoration work being carried out on the fresco paintings and rock structures. The 'cult of history' and the nostalgic yearning in tourism is very much evidenced here, both in the designation of the Göreme valley as a museum and in its being considered of *world* heritage significance. Indeed, almost all of the tourists who visit the Cappadocia region do visit the Göreme Open-Air Museum in order to view the well-preserved caved Byzantine churches within.

The Byzantine churches are, of course, the heritage of the Christian world, and so, through the emphasis placed on the importance of these sites in tourist representations of Cappadocia, the more recent Islamic settlement in the region has in turn become de-emphasised (see Tucker 2001). Historically, Islamic Turks have inhabited the Cappadocia area since the twelfth/thirteenth centuries, following the defeat of the defending Christians by the Seljuk army and other Turkic tribes.[3] The site of the village named Göreme today, approximately two

Plate 2.3 The coach park by the Göreme Open-Air Museum.

kilometres from the Byzantine monastic settlement of Göreme, was settled as a farming village called Maçan. The oldest mosque in the village is dated 1686. Following the founding of the Turkish republic, the village was given the more modern Turkish name of Avcılar, meaning 'Hunters'. In the 1970s it was renamed Göreme, this time by the municipality itself, so that the village would be associated more directly with the growing fame of the nearby Göreme Open-Air Museum, and also so that it would become eligible to receive part of the museum's earnings.[4]

The association of the village with the monastic museum site has resulted in a certain level of confusion as to which identity should be associated with the caves in the area. A 1970 article in *National Geographic*, for example, says:

> We had come to learn more about the ancient Christians who had carved these caves. During the month ahead, we would try to live as they had lived.
>
> (Blair 1970: 127)

Although the American couple writing this article undoubtedly learned the way of life of the Muslim Turks living in the caves more than that of the earlier Christians, they still insisted on representing the caves through a Christian idiom:

> **Muslim Prayer From a Christian Cave:** Until age had weakened his voice, Yusuf had been a muezzin, whose duty it had been to call the Muslim faithful to their prayers five times daily from the mosque minaret. With Ayşe interpreting, I asked him if he would mind calling the prayer from a Christian cave. He could see no harm in it. I quietly flicked on my tape recorder as he began with a quavering voice:
>
> Allah is the highest. I am a witness that Allah is one.
>
> Mohammed is his Prophet. Come to prayer, He will give you comfort...
>
> (Blair 1970: 138)

Similarly, this apparent confusion of the Christian past with the Turkish/Islamic present in Göreme is shown in the following tourist's answer to my asking what had interested her in coming to Cappadocia:

> My boss had been here, and then I saw pictures. And I studied ancient history so I was really interested in the Christian parts in the caves. I've really never seen anything like this before, it's a much more primitive society. My relatives had been here and they raved about it, and also I'd seen pictures. It's just surreal the way the fairy chimneys look – it's just like a Dali painting. And it's the fact that people lived in it – all the little doors

and windows. I didn't even know about the frescoes before I came and that is not all that interesting to me. It's a kind of living history. The people are still here, they are still living in the caves – and there are bars and *pansiyons* in the caves. You see, we're from Australia and so we come here to see the history.

The constant re-hashing of the images presented in updated versions of the travel literature, always providing great detail about Göreme's churches, perpetuates the tourist myth of Cappadocia as being a 'Christian place'. The local inhabitants of the region have also now appropriated this myth, believing that tourists come to the region in order to 'see the graves of their ancestors'. Stemming from this belief, as well no doubt as from the potential tourist income they can derive from tourists in the area, is a general adoption by the Göreme villagers of views concerning the importance of the preservation of the churches and their frescoes:

> For us, Göreme has the most significant wonders of Cappadocia. It is an epic which the rain has shaped together with the wind and written into the nature. With its fairy chimneys which cannot be seen in any other part of the world and its churches carved of rocks which represent the most interesting pieces of Christianity and its monasteries, Göreme is a treasure of history and culture.
>
> (Mustafa Mızrak, Mayor of Göreme – Mızrak 1995: 5)

When asked in interviews why they thought tourists come to the area, villagers often answer, 'for pilgrimage, just like we [Muslims] go to Mecca'. In this view, the villagers' own role in Göreme's tourism is simply as providers of the services necessary to enable tourists to stay in the area and see the churches and the landscape:

> Göreme, a small township hidden in fairy chimneys, is also called 'the heart of Cappadocia.' With its 2000 permanent residents, Göreme is...a great historical and tourism centre....The people of Göreme are exerting their utmost efforts to host their foreign guests.
>
> (ibid.)

Apart from asserting their position as 'hosts' to tourists, Turkish-produced tourism promotion literature tends largely to ignore the presence of the Turkish villagers living in the caves today. This is due both to their following of the myth that it is the Christian historical remains that tourists travel to Cappadocia to see and also, understandably, because of their cultural distance from the Romantic 'cult of the primitive' that cuts through Western modernity and tourism.[5] Many representations of Cappadocia in the foreign travel literature and tourism promotions, on the other hand, do highlight images of the villagers' rural and 'troglodyte' way of life.

Göreme as a cultural landscape

> Even today many of the caves are in use; Turkish farmers live in some, and others serve as stables and storerooms. New cave houses are still occasionally hacked out by peasants who prefer them to more expensive conventional homes...Not infrequently, Cappadocians dwelling in cones will warmly invite tourists to 'come inside and see our home'.
>
> (Blair 1970: 129, 146)

> Even today many of these caves and grottoes serve as homes and store houses for peasant families. Whole villages of cave-dwellers still exist.
>
> (Explore Worldwide 1989)

> Thousands of years after the Stone Age passed into history, here in the extraordinary landscape of the Cappadocia region cave life is considered healthy, economical and even chic.
>
> (Kinzer 1997)

> Göreme...still has a thriving community working the fields tucked away between the fairy chimneys and carrying on community seasonal activities such as autumn harvest of pumpkin seeds and the preparation of 'pekmez' and 'village bread' to see them through the long winter months.
>
> (turizm.net 2002)

It was noted above that the 'cult of the primitive' (Mumford 1934, cited in Wang 2000) is the Romantic reaction to modernity, creating an aesthetic valuing of a simple or natural life associated with pre-modern times. This yearning lies behind the 'ethnic' or cultural forms of tourism that encompass a tourist search for people who appear to be living pre-modern lives. A similar idea was put forward by MacCannell (1976), who argued that tourists are alienated by the conditions of contemporary life and thus, in an attempt to recreate the structures that life in the modern world appears to have demolished, search for authenticity in other times and other places. The representations of Cappadocia quoted above clearly follow this Romantic search for a simple and pre-modern world. This search is met in images of a 'thriving community working the fields' and making 'village bread', and it is met most conclusively in the notion of a people still living in caves.

Those people who inhabit the caves and rock structures thus become part of the 'extraordinary' landscape itself, so that there develops a cultural landscape that is aesthetically valued and appropriated for tourist consumption. This process is shown in the following extract from a leaflet prepared by the Göreme National Park group in the mid-1980s:

> The Göreme Historical National Park shall be protected and developed, so that the present and future generation can benefit from the scientific and aesthetic nature, as well as the natural and cultural values...The picturesque

village life, the activities of the villagers, the small volcanic farming areas...All these peculiarities, the tufa rocks and fairy chimneys as they are in traditional relations, are adding to a moving and vivid view...The preservation of this traditional view is the main theme of the administration, protection, presentation, and the development of this historical National Park. At the application of the National Park, the main policy has been adopted that the population living within the boundaries of the park should be one of the main important elements, as well as giving support to the resources.

Although this National Park leaflet is produced in Turkey, the content is drawn from a report written by a group of national park advisers from the USA employed by the Turkish government in the 1960s to visit and provide advice regarding the development of the Göreme National Park. Because that group of Americans considered the 'culture' embedded within the Cappadocia landscape to be eligible for tourist attention, they included it along with the fairy chimneys and Byzantine churches as part of the park's resources and thus worthy of protection and preservation.

These images serve also to reinforce 'shared understandings of cultural difference' (Lutz and Collins 1993: 2), immediately positing the cave-dwelling inhabitants of Cappadocia as an 'other' people. This in turn leads tourist representations to emphasise only those elements of 'other' cultures that are different to those of the tourist's culture (Said 1978; Buzard 1993). It thus becomes clear why the foreign-produced images of Cappadocia are more likely than the locally produced images to highlight the 'troglodyte peasant' culture as an attraction of Cappadocia. Western travel writers and guides, equipped with a Romantic tourist gaze, are more likely to recognise the points of difference with their own culture and to aesthetically value and promote them.

Repeatedly seeing the visiting tourists' fascination with the caves, however, the people of Göreme have gradually come to appreciate the value of the caves and the opportunity to sell tourists the chance to become cave-dwellers themselves. So while the representations of the village in promotional material produced by villagers tends not to dwell on their own 'troglodyte peasant' identity, it promotes the tourists' opportunity to sleep in a cave, to drink in a cave-bar and to eat 'traditional', 'home-made' food. Advertisements for Göreme's *pansiyons* in tourism promotion literature, in the accommodation office in central Göreme and now also on the Internet, highlight their 'traditional' cave rooms and their breakfast-terraces overlooking views of the village and the fairy chimneys:

Saksağan Cave Hotel is an old house which, having been restored, becomes your own home in Göreme, with rooms that offer you history, nature and culture together, Saksağan Hotel provides a super view overlooking Göreme. Rooms are located within fairy chimneys and carved into rocks.

(wec-net 2002)

Today, with numerous such tourist accommodation establishments in Göreme, tourists can, for a time, become troglodytes and a part of the cultural landscape themselves.

Touring Cappadocia

During the last two decades Cappadocia has increasingly been constructed and represented as a tourist landscape. Tourists are told in representations of the region not only about the extraordinary natural landscape and the historic and contemporary culture embedded within it, but also of the tourism services available to them. Along with the accommodation, restaurants and shops, the whole region, itself a mythical construction for tourism,[6] is presented in tourism literature as a collection of sites that are there for the purposes of tourist consumption. Besides the Göreme Open-Air Museum, the main sites advertised to tourists include two 'underground cities' at Kaymaklı and Derinkyuyu, situated approximately fifty kilometres from Göreme and comprising deep networks of underground tunnels and dwellings in which early Christian populations are said to have taken shelter from invading Arab armies. Another site is the Zelve Open-Air Museum, which is situated seven kilometres from Göreme village and is similarly presented as a Christian site even though the area was inhabited by Muslim Turks until rock collapse gave rise to their evacuation in the 1950s.

Tours of the region are set up around the tourist map, which creates an imaginary boundary around Cappadocia and an imaginary geographical structure within that boundary, so that tourists and local people alike – those local people who work in tourism at least – have the region and all its tourist sites and paths between those sites mapped out in their minds. Having read tourist literature myself when I began my fieldwork in Göreme, I too had this tourist map of Cappadocia in my mind, and the map became even more pronounced while I helped out in Abbas's tour agency advising tourists on what to do in the area. All of the agencies in Göreme have a large map of the Cappadocia region painted on a wall and I constantly referred to such a map as I described what was on offer to tourists.[7] My geographical sense of the region became heavily structured around the day tour routes and where the main tourist sites are in relation to each other (see Figure 2.1).

Almost all of the tourists visiting Cappadocia go to the Göreme Open-Air Museum and one of the 'underground cities', as those are the main iconic sites of the region. The practices and experiences among the tourists vary greatly, however, dependent particularly on whether or not they are visiting Cappadocia with a package bus group tour. The package tours to Cappadocia all follow much the same pattern. Tourists either travel via an international travel agency that has fully handled their travel arrangements from their departure until their return home, or they purchase a three-day tour of Cappadocia while on holiday either in Istanbul or on Turkey's south-west coast. On such tours their agency places them in one of the larger, higher-rate hotels in the region, most of which are situated

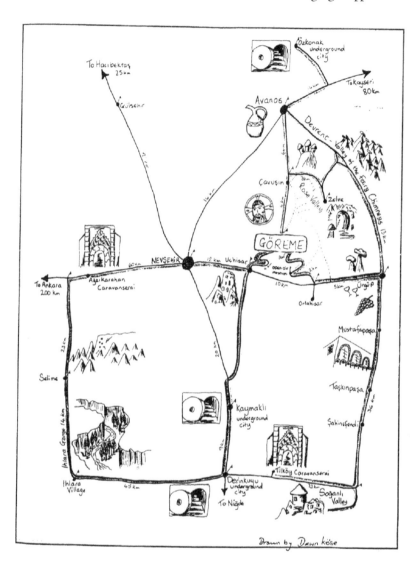

Figure 2.1 Tourist map of Cappadocia showing the main tourist sites.

outside the boundary of the Göreme National Park, either in Ürgüp or on the edge of Nevşehir. Their few days in Cappadocia have a full itinerary: getting up and leaving the hotel early each morning; bussing around the museum sites and tourist shops; and then being entertained in the evening either at one of the region's folk-dancing shows or by a similar programme organised at their hotel.

Tourists who are on group tours are invariably placed into a situation of what Bruner calls 'touristic surrender':

> Touristic surrender involves acceptance of the common practices of the
> group tour, such as the social requirements of group travel and the loss of
> the ability to set one's own agenda.
>
> (Bruner 1995: 237)

Although there may at times be scepticism or even dissent among tourists on a
group tour, the practice and experience of Cappadocia for bus group tourists is
heavily structured by the agency and guide. Throughout their Cappadocia experience they are accompanied by a nationally certified tour guide who is
invariably from outside of the region, often from Istanbul. The guide shares
with the agency control over the tourists' viewing and experience of
Cappadocia, and so the tourists are also led to surrender to the images and
myths that I have outlined in this chapter because it is those images and myths
that the agencies and guides employ and perpetuate in their own practices. This
point is made by Weightman in reference to package tours in India,

> false impressions generated by distorted imagery in tourist literature are
> compounded by the content of the package tour brochure as well as the
> tour itself.
>
> (Weightman 1987: 229)

To view the lunar landscape the tourists are taken to signposted 'panoramic
viewpoints' on high points overlooking cone-filled valleys, or to clusters of fairy
chimneys that have been highlighted by photography and descriptions in the
tourism literature. These quick landscape-viewing stops usually take place *en
route* to specific sites, such as the open-air museums and underground cities,
where the guide might give a lengthy explanation about the presence of
Christianity in the area and life in the caves during the Byzantine era. On some
tours guides take their group to visit a contemporary family's cave house. Here
the guide acts as mediator between their group and the cave-dwellers, representing the latter in a way that fulfils the tourist image of the 'troglodyte
peasants' outlined above (this is discussed further in Chapter 6).

In the situation of 'touristic surrender', therefore, 'tourists also surrender
control of their relationships with the [local] people' (Bruner 1995: 237).
Apart from the occasional commercial transaction conducted at souvenir stands
at the museum entrances, tourists on package group tours have very little or no
contact with local people. They are presented in the brochures with images of
the people of Central Anatolia being 'the most hospitable people, proud of
their heritage' (Tapestry Holidays 1997), yet in their hotels and at the folklore
shows the staff are as likely to have come from other parts of Turkey where
they were trained in tourism-service programmes. Guides are keen not to let
their group roam freely in villages such as Göreme, trying to prevent them
from buying anything until they reach the bigger tourist shops that are
purposefully built for bus groups of tourists and situated on main roads outside
the towns and villages. The guides have arrangements with these shops

whereby they receive a hefty commission on anything sold to their tourists. It thus becomes clear how tourist images and myths are entwined with the commercial aspects of tourism.

The tour agencies and guides have significant control over the practices and experiences of their tourist groups. It is this level of control that is considered by some tourists, and in some tourist representations of Cappadocia, to be problematic, as this guidebook to Turkey illustrates:

> Bear in mind, however, there are distinct problems with guided tours that are often compounded in Cappadocia. Although the guided tours visit many of the area's most interesting sites, they don't give you the time to get off the well-worn paths and poke around in out of the way areas, or climb through a warren of caves to the top of a cliff and rest…Second, guides take you on a circuit that can get congested, and most guides are inclined only to show you what they want to show you. You may stand in line while 200 feet away a collection of churches sits empty. Third, the amount of time spent visiting the ruins can be equivalent to, or even less than, the amount of time at carpet, pottery, and rug-weaving shops, and these establishments are no great bargain. The prices aren't especially good, even before the shop-owner tacks on the commission he is turning over to your tour company.
>
> (*Open Road Publishing Turkey Guide* 1996: 288)

Much of the tourist literature on Cappadocia, particularly that aimed at the 'independent' tourist market, suggests in this way that a somehow truer experience of the region can be had by getting off the main tourist paths. This idea appeals to an 'alternative' tourist gaze, as mentioned at the beginning of this chapter. Behind the alternative gaze in part is a strong tourist concern about the 'polluting' influence that tourism and tourists have on landscapes. In *The Good Tourist Guide to Turkey* (Wood and House 1993), for example, it is stated with regret that Göreme village has been transformed with the building of a 'hideous' tourist centre. A further example comes from the *AA Essential Explorer* guidebook to Turkey, in its description of the Göreme Open-Air Museum site:

> The churches of the Göreme Valley contain unquestionably the most spectacular paintings of the region; as a result, this area has been designated the Göreme National Park, a huge open-air museum. The inevitable corollary is heavy development of the site for tourists, with high entry and parking fees, a bank and shops at the entrance, and a definite sense of being 'processed' round, as you follow the arrows in a one-way system.
>
> (AA 1995: 176)

This description of the museum is warning tourists against being caught in the same uniform gaze with the rest of the tourists, all following the arrows and looking the same way. This is an alternative tourist discourse that promotes the ideas of discovery and adventure, and produces a representation of Cappadocia

for the alternative gaze: 'For those who take the time to explore the less travelled byways, Cappadocia is a land of discovery' (Adventures Great and Small
2002). Many tourism promotion texts point out that sections of Cappadocia
remain unpolluted by tourism and therefore still available for those with a more
discerning judgement and taste. *The Good Tourist Guide to Turkey*, for example,
points out that:

> Despite what could sound like a tourism horror story, Cappadocia is big
> enough to absorb most of its visitors without too much pain. It doesn't
> take much effort to get away from the herd, and there are still lots of
> outlying sites which receive hardly any visitors at all.
>
> (Wood and House 1993: 236)

Some representations of the region, then, point the way for a second tourist
gaze, one that attempts to resist the primary, and apparently obligatory, way of
seeing Cappadocia. To some extent, all tourists' practices and experiences in the
Cappadocia region are guided by the tourist images outlined above. They are all
chasing the tourist myths that have constructed Cappadocia as a tourist place,
and following representations of the lunar landscape and the cultural landscape
that encompasses both the remnants of a Christian history in the area and the
rural troglodyte Turks who live in the caves today. However, tourists might also
have a desire to see beyond those tourist representations, and some more than
others are able to do so. They are the ones who are not on a package group

Plate 2.4 A hotel built to resemble the 'fairy chimneys' of Cappadocia. This hotel is
situated outside the Göreme National Park near Nevşehir.

Plate 2.5 The kind of vernacular accommodation that backpackers prefer to stay in.

tour that follows a preset itinerary and has a guide who attempts to control not only their movements and experiences but also their interactions with local people they may meet.

In contrast to the tourists who come to Cappadocia on a bus group tour, most of the independent tourists visiting Cappadocia stay inside towns or villages in smaller, locally run *pansiyons* or hotels. They do sometimes take day tours of the main sites in the region offered by local tour agencies, and most of them also follow the idea that the Göreme Open-Air Museum site is the main place of touristic interest in the region. However, they also spend a lot of their time in Cappadocia following the discourse of discovery and adventure, walking in the valleys that surround Göreme village and exploring and clambering in and out of caves and churches that they like to discover for themselves. Much of their time is also spent simply hanging out in the village in which they are staying: eating, drinking and shopping at *their own* will, and sometimes wandering through the back streets to take a look at an 'other' side of Anatolian, and in this village 'troglodyte', life.

In short, though they are like the package tour tourists to the extent that they are also chasing the tourist images of Cappadocia in their coming to the region in the first place, independent tourists are freer to choose the extent to which they partake or do not partake in the chasing of the images and myths while they are there. So while they are by no means *off* the 'tourist path', nor eluding those images and myths altogether, they are to some degree at least free to look the other way, or indeed to not look at those images at all. They engage fully in what MacCannell (2001) has termed the 'second gaze', and a great many visitors to Cappadocia follow this gaze.

To summarise, Cappadocia has been constructed as a tourist place through the production of certain images and myths, initiated through earlier travel literature and then cultivated by the tourism and heritage industries. These representations feed off Romantic notions of extraordinary and pristine land-scapes, a monumental past, and the innocent simplicity of a pre-modern life, as well, in part, as orientalist demarcations of what is different between 'us' and the 'other'. Earlier representations of the Cappadocia region thus marked the 'lunar landscape', the Byzantine history and the 'troglodyte peasants', and these have been duplicated in a variety of promotional contexts: in brochures, guide-books and websites, all denoting what *should be* of interest to tourists and what their gaze *should be* directed towards.

Many tourists visiting the area, however, and particularly those travelling independently of an organised packaged tour, try to look beyond the key tourist representations with which they are presented. MacCannell has recently suggested that this second tourist gaze involves 'a drive to provide narrative accompaniment to what is seen' that 'always goes beyond descriptions of the visible' (MacCannell 2001: 32). To look at the primary tourist gaze on Cappadocia, as I have in this chapter, therefore tells only part of the story, the part manufactured by the tourism industry. To understand the whole gaze and experience of the actual tourists visiting Cappadocia, it is necessary to look at the ways that tourists both construct and resist that primary gaze through their narratives of identity and experience. It is to a more detailed account of that alternative, or actual, tourist gaze that I turn in the next chapter.

Notes

1   These figures were taken from the numbers entering the Göreme Open-Air museum, and are based on the assumption that almost all of the tourists in Cappadocia visit this museum.

2   See, for example, Haber (1995), Hirsch and Hanlon (1995), Urry (1992) and Wang (2000).

3   Until the Greek/Turkish population exchange in 1923, prompted by Ataturk and the founding of the Turkish Republic at that time, many villages and towns in the Cappadocia region were most likely inhabited by both Christians and Muslims.

4   Although now, since a change in the national law concerning historic sites and the total appropriation of the museum by the National Ministry of Culture, Göreme village receives no direct income from the museum ticket sales.

5   See Tucker (2001) for discussion on these issues in relation to Zelve Open-Air Museum, situated approximately ten kilometres from Göreme.

6   The name 'Cappadocia' does not really exist, either geographically or socially, in modern Turkey, but is an ancient name for the region dating, according to most accounts, from the time of the Roman Empire.

7   A similar process is described by Black (1996) in talking about her work as an infor-mation officer for the London Tourist Board.

3 The tourists

In search of serendipity

We are tourists but not typical ones. We try not to be, but we are. We try to go where there isn't so much tourism. We wanted to see Göreme but we only spent one night here. We are not staying in the Tourist Hotel or anywhere like that, but we're also not meeting people like staying with some man in the middle of nowhere. We have a programme, like we try to see things like the Aya Sophia and the Göreme museum. The real tourists are going on buses, stopping, looking, drinking Coca Cola – whereas we spent two hours climbing in Zelve – so we're not really like tourists. We try not to be typical tourists, but still, we are. In two weeks you can't see the way they live, and we've always had a European toilet!...We are on the outside. Turks here live on tourism, so they look at us as tourists, to get our money – not that they think of money as much as we do.

(Belgian tourist in Göreme)

One of the things I like best about travelling is getting up in the morning and having no idea who I'm going to meet that day, or what I'm going to experience. Some of them may not be particularly pleasant either, but they are memorable.

(American tourist in Göreme)

These extracts from interviews with tourists in Göreme show that the identity, experience and gaze of tourists are not necessarily as structured and set in a particular mould as they are so often assumed to be. This is because, as suggested at the end of the previous chapter, their gaze and experience can go beyond the tourist representations laid out for them. The key point here is that tourists do not merely engage in a visual, directed and distanced gaze, but they also *interact* with the places and peoples they visit, as well as with each other. The world in which tourists move, therefore, is also active in the writing of their experience, and so that experience should be seen as always under negotiation with the people and places they meet.

It is necessary to consider this process of negotiation in order to reach a full understanding of how the experience and gaze of the tourists in Göreme is constructed. Having described the tourist images and myths surrounding Cappadocia, or the primary tourist gaze, in the previous chapter, this chapter will focus on another gaze, one that is not pre-determined and scripted prior to the tourist's visit.[1] This other gaze is produced and negotiated in interaction during the trip through the narratives the tourists construct, and thus depends to a large extent on what *happens* during the trip and on what and who each tourist encounters along the way. The aim of this chapter is to elucidate this other gaze by looking closely at the travel 'style' of the tourists in Göreme and the ways in which that style is constructed through their travel experiences and narratives. First, I will give a broad overview of the tourists and their travel habits. I will then outline portraits of a sample of tourists, before going on to elaborate the key themes in their travel narratives.

Tourists in Göreme

Most of the tourists who stay in Göreme are travelling independently of organised tour groups, and almost all of them are international tourists. They generally either come from western Europe (including France, Germany, the Netherlands, the UK and Italy) or are English speakers from Australia, New Zealand or the USA. Increasing numbers of young backpackers from Japan, Hong Kong and Singapore also began to visit Göreme during the 1990s, as well as a small but growing number of independent travellers from eastern Europe.

They usually enter Turkey by flying into Istanbul, though some do travel over land or sea from Greece, and most of them are 'backpacking' a circuit from Istanbul to Cappadocia, and then on to the south-west coast of the country from where they make their way back up to Istanbul. The Australians and New Zealanders usually stop at Çanakkale on the west coast to visit the site of Gallipoli, where Allied soldiers, mainly Australians, New Zealanders and British, were slaughtered by the Ottoman Turks during World War I. Many tourists go to that site to attend the memorial service held on 'ANZAC Day' in late April, and so in Göreme business from Australians and New Zealanders picks up around that time. Few of the tourists go further east from Göreme, as it is generally considered that once you go beyond the Cappadocia region you are entering more dangerous ground, particularly regarding the Kurdish troubles in the south-east and east of the country.

Very few Turkish tourists stay within Göreme, though with the increasing popularity of Cappadocia as a region of cultural interest, an increasing number of Turks do come to the region to visit the museums. They generally stay in the bigger hotels in nearby towns such as Ürgüp, however, along with the many other 'cultural tourists' of various nationalities who visit the region on bus tours. In addition, at the weekends, particularly in springtime before the summer vacations begin in Turkey, the region fills with busloads of Turkish school parties coming to visit the historic sites. Although again these tourists do not stay within Göreme village, their presence in the area is strongly felt on Saturdays and Sundays because of the huge increase in traffic moving through the village centre. On some days the situation could only be described as one of chaos. One Sunday in May, for example, I counted as many as thirty coaches in the car park of the Göreme Open-Air Museum in the mid-morning; all of those probably drove through the centre of Göreme village earlier that morning. The traffic problem created by these buses is coming to be considered a serious one, and during the two years of my field-work the municipality made extensive alterations to the main road by installing speed bumps and a central reservation, and by attempting, somewhat in vain, to forbid villagers from driving their donkeys and carts on the main road.

The international backpacker tourists in Göreme usually carry a guidebook, the most popular of which are the *Lonely Planet* or *Rough Guide* for English speakers and *Le Guide du Routard* for the French. Besides the advice on where to go, what to do and see, plus a limited amount of information about history and culture, these guidebooks provide advice on where to stay, eat and drink. There conse-quently tend to be enclaves of certain nationalities in different *pansiyons*. The tourists also follow advice from other backpackers they meet all around Turkey, and so many of them travel a similar route around the west half of the country and stay in the same places. Most of them move around the country using the public bus services, which are generally cheap and efficient. They usually stay in Turkey for a length of time between two and four weeks, and the time they stay in any one place, including Göreme, usually varies between three and ten days. Some of the tourists, especially those from Australia and New Zealand, are on extended trips from home, many lasting between one and three years. Most of the Australians and New Zealanders have working-holiday visa entry to Britain and are using Britain as a base from which to travel throughout Europe, the Middle East and Africa. Some of those with no particular time limit stop and work in Göreme's tourism businesses for a few weeks or even months. Some return for the next season, and a further few take up residency in the village.

Some tourist portraits

Jake and Susan

Both in their late twenties, this Australian couple had been travelling for eight months around the world by the time they arrived in Göreme. They were taking time out between university and getting a job, something they said is very normal nowadays with Australians: 'If you haven't done it, there's something wrong with you.'

Jake had been to Göreme before, in 1989, and felt pleased that it had not changed much. He came mostly for the landscape and the caves: 'You have to go into a cave to experience it for yourself.' Before he came the first time, Jake had not known about the Byzantine churches in the area, so they certainly were not a factor in his initial motivation for coming to Göreme. The couple also enjoyed Göreme being a 'traditional' village: 'It's nice for time to stand still in some places,' said Susan. While in Göreme they chose to stay in a small family cave *pansiyon* situated in the back streets of the village. They ate the Turkish breakfast and the evening meal served in the *pansiyon*, and most days went out 'exploring' in the valleys. They also visited the Göreme Open-Air Museum and hired a local guide there to explain the history of the churches. They did not consider themselves 'tour people', however, and felt uncomfortable about the pressure they seemed to be under to book a day tour from a Göreme tour agency. They thought it better to 'discover it all for yourself'.

Max

A German masseur of approximately thirty years of age, Max was travelling alone around Turkey for two weeks. He had decided to take a break from work and had come to Turkey because that was where his travel agent had some cheap flights available. The travel agent had also told him about this famous and 'mystical' place called Cappadocia in Central Turkey, with a fantastic landscape, cave-houses and underground cities. While staying in Göreme, he had visited the Göreme Open-Air museum and made his own way to an underground city by taking two public buses. He said that he would not take any day tours because he preferred 'to be free to being guided and controlled'. He said that he preferred to explore in the caves and climb up tunnels whenever he wanted to, though he did tag onto the back of a tour group in the Open-Air Museum in order to listen to the guide's explanations about the symbols of the frescoes in the churches.

He remarked that when he travels he likes to learn about the people; that:

> in a foreign country I try not to be like a real tourist, I try to be very careful, even though it is like walking on a knife-edge sometimes. I can't be Turkish, so I have to remember that I'm the guest.

When asked what he considered 'real tourists' to be like, he replied that they go around in luxury, that they are always on tour and have nothing to do with the local people nor do they respect the local culture, and they 'just want to eat German food all of the time'. He, conversely, does try to respect the local culture; he was always careful in Turkey not to sit next to women on buses, and he did not walk around without a T-shirt on.

He thought that it was good for the people in Göreme to have tourism because it allows them to get in contact with the Western way of life, thus widening their horizons. However, 'tourism can destroy the culture', because it 'introduces money and competition'. 'Here it is a communal life, the people are happy and social, even though it must be hard work here', but 'tourism is destroying the families', it 'causes people to fight for a living', and 'money makes people crazy sometimes'.

Max expressed a lack of ease where 'money' issues were concerned in his tourism experiences. He found himself being distrustful in Turkey because it is such a different culture. At first he thought he really had to be careful in case he was ripped off, but he found it to be the opposite. 'People are far more friendly and generous than I had expected; I suppose it's in their culture.'

Julie and Sarah

From Britain and New Zealand respectively, and in their early twenties, this was the first time that Julie and Sarah had visited Turkey and they were on a two-week 'backpacking holiday'. They had become friends in London where they both worked for the same pharmaceutical company. They chose to come to Turkey because it is a cheap country to visit, has warm weather and seemed interesting. Julie said that they thought it would 'widen their horizons', but then she laughed at herself for coming out with such a 'cliché'. They thought, though, that it would be an interesting challenge to travel as two women in a country where they did not know the language, and which was considered at home to be mildly dangerous. They had received lots of warnings to be careful in Turkey: 'My mum thinks I'm crazy coming here', said the Londoner, 'but the more I got told, the more determined I became to really come here.'

They came to Göreme by public bus from Istanbul, and were spending about four days there before travelling down to the south coast and gradually making their way back to Istanbul. They wanted to come to Cappadocia first because they had seen pictures of the rock houses. They had chosen to stay in Göreme village following their guidebook's advice, and also because they had noticed that all the other backpackers had got off the Istanbul bus in Göreme. They were pleasantly surprised on arrival in Göreme that the area of cave houses was much bigger than they had expected and also that the caves were still inhabited. They saw Göreme as being very much a 'tourist-oriented' village, however, since 'everywhere there is either a carpet shop, a *pansiyon*, or a tour agency', and 'everyone speaks English'. They also found it strange that there were 'no women around'.

While in Göreme they spent a fair bit of time relaxing in the sun on the roof of the cave *pansiyon* they had chosen to stay in from the many that were advertised in the 'Accommodation Office' in the bus station when they had arrived in the village. Also they had visited the main tourist sites: the Göreme and Zelve Open-Air Museums, and also an 'underground city' that was part of the itinerary of a day tour they arranged from one of the agencies in the village. They has also sampled the night-life of Göreme, visiting one of the bars and the 'Escape' cave-disco, where they enjoyed dancing to the English pop music.

The two women reflected that their travelling mentalities were different from each other because of their different nationalities. Julie had only ever been on package holidays before, but this time decided to travel in a less-organised manner because she wanted to be able to relax between sightseeing. She did not want to sightsee all of the time because it becomes a bit of a blur, and 'after you've seen a few of the churches, it's enough'. Sarah, on the other hand, felt more of a sense of urgency to 'see everything' because Turkey is so far from New Zealand and she might never come back again. That is why, she said, New Zealanders tend to 'really backpack around'.

Akira

Coming from Japan, Akira had been travelling for three months by the time he came to Göreme. He had made his way overland from China following the 'Silk Road', and was particularly interested in coming to Turkey because of its position 'between Asia and Europe'. In his mid-twenties, he had always had a passion to travel abroad because he felt Japan to be a 'very systematic country' and he wanted to 'be released from that system'. In doing so, he felt that his travels would develop his character:

> If I work for a Japanese company for a long time, my thinking becomes very narrow. But now I've travelled, I've got other ways of thinking.

He noted that it was unusual for Japanese to travel outside of organised tour groups, and everyone at home had told him it would be dangerous to do so.

Akira came to Göreme because he was 'interested in the strange view' he had seen in pictures of the landscape. He had also wanted to come to somewhere rural. He was enjoying the experience of staying in a cave *pansiyon* for a few days, though he said that he would not like to live in a cave all the time because it was not actually very comfortable. He found it amazing that people do live in caves there; that they are 'living with and coping with nature'. He felt that although Japan was economically superior to Turkey, Turkey was superior in terms of the way of life of its people:

> Japanese people perhaps used to be like Turkish people, but the system is now complete in Japan, and time is money. In Turkey, there is lots of time…Maybe I want to change, but I can't in Japan – but I want to remember this life in Turkey, and then when I retire, when I'm sixty, I want to live like Turkish people. Turkish people get up with the sun and sleep with the sun, it's a very slow life.

During his time in Göreme, Akira visited most of the main tourist sites, walking to the Göreme and Zelve Open-Air Museums, and he also spent time resting in the garden of his *pansiyon* playing backgammon with other travellers. He had walked around the village a fair bit, both in the centre and up into the older winding residential streets. Up there he enjoyed the feeling that he was really a foreigner, and 'liked passing and smiling at people'.

Dan

A furniture maker from the USA, Dan was travelling around Turkey for a few weeks. He had stayed for two nights in the nearby town of Ürgüp, but came to Göreme because in Ürgüp he felt uncomfortable because of all the tourist shops there: 'I don't like walking around with everyone looking at me as if I'm a walking dollar.' He saw both Göreme and Ürgüp as 'artificial', and resented 'all this tourist stuff here'. He preferred to see himself as a 'traveller' rather than 'tourist', and described tourists as those who stay in three-star hotels, expect high services and to be entertained, and 'don't have proper interaction with the local community'.

During his search for cheap accommodation in Göreme he became annoyed that he was unable to bargain the price of a room down. He had eventually managed to bargain for a room in a *pansiyon* that was actually closed because it was a cold November and this *pansiyon* had no heating. He got a bed for US$3.50, down from the usual US$5 rate for beds in the village. He kept asking me how much 'locals' would pay, as he hated the idea of paying more than a Turk. I told him that the prices in Göreme were cheap because they were aimed at backpackers and that Turks would usually pay more if they went on holiday. He argued that if he was prepared to rough it then he should be able to sleep on the roof for a dollar. He thought that the best way to travel was to be adventurous and daring, and in Turkey he had been to the south-east and 'loved the PKK areas [areas where fighting was taking place between the Kurdistan Workers' Party (PKK) and Turkish military] more than anything'.

Group of Australians

I met this group of around ten Australians one morning as I was cycling through the village and they stopped me to ask where I had rented my mountain bike and how much for. Having just arrived that morning on an overnight bus from Antaliya on the south coast, they were sussing out the scene in Göreme and seeing what was available for them to do. They had all come to Turkey for a few weeks backpacking from London where they were living on working-holiday visas. They had started in ones and twos but, because they all followed the Lonely Planet guidebook, they kept meeting in the same towns and *pansiyons* around the coast. Gradually they had formed into a large group, both male and female, all in their early to mid-twenties, and had travelled up to Cappadocia on the same bus from the south coast.

They seemed to find it quite amusing that I had been living in Göreme for an extended period of time, but they were pleased to have someone who could recommend things they should do and see during their stay there. They had heard about the Flintstones Bar from other backpackers at the south coast, and they asked me where it was, what it was like and what time it would open that day. They then concluded that they would spend the rest of that day in bars, and then do the day tour to the underground city the next day before they got on the night bus to Istanbul the following evening.

They asked me to recommend a good but cheap restaurant where they could get breakfast so I took them to a restaurant and sat with them for a while. Though the waiters there were very friendly with the group, if a bit sarcastic in their putting on Australian accents to take their orders, the

Australians did not seem at all interested in interacting with them. They seemed more interested in chattering with me and with each other. After breakfast, as we parted company, they said they hoped to meet me again in the Flintstones Bar later that afternoon.

Alison and Clare

While working in England with a working-holiday visa, these Australian friends had come backpacking to Turkey and had both ended up finding boyfriends in Göreme. They had been staying in Göreme off and on over the past year between going back to London periodically during the winter months to earn some money. They had previously had no intention of staying in Göreme: 'I came as a tourist and thought I'd keep going…and I did, but I came back, and then I stayed.'

They had both come to Turkey on a month-long backpacking holiday, along with thousands of other young Australians and New Zealanders who come for the 'ANZAC Day' celebrations at Çanakkale. They were combining this remembrance trip with the usual backpacking circuit around the west half of Turkey. When they came to Göreme, however, they had fallen in love with Göreme men and were now deciding to stay and 'make a life' with these men in the village.

I asked the women what had motivated them to set off travelling from Australia in the first place:

Because I'd had enough of normal life, working, paying for the car, going out every weekend, and I thought it was time to get out and see the world…so I did.

The other one added:

I was a bit lost at home, so this was my time to take off and…I'd come out of a relationship, and then my home environment wasn't quite the same because the relationship wasn't there anymore. And I'm not career-orientated and I just didn't really know what I was doing, and I was just sort of living day by day, week by week…and then Alison said 'let's go away', and I said 'yeah, let's go'.

Both of the women had been working in the *pansiyons* and tour agencies of their boyfriends while in Göreme. They both felt that by staying there, they had somehow escaped their humdrum lives at home:

I'd rather stay here than go home – home is where the real people are. They've got real lives, and they've got real jobs, and they've got real mortgages. Göreme is just like a fantasy. It's just not a rat-race here – it's excellent, it's really me!

Tourists, non-tourists and the ill-fated plot

Underwriting all of these tourists' narratives is the distinction between 'tourist' and 'traveller' and a deliberate will to assert a traveller over a tourist identity. The ways in which some tourists attempt to shake off a tourist identity has been discussed in a number of academic publications, revealing in a variety of ways the problematic nature of that identity.[2] Riley (1988) and Dann (1999), for example, have both discussed ways in which travellers separate themselves spatially and temporally from tourists. Elsrud (2001) has looked at how the grand narrative of risk and adventure running through backpackers' travel stories forms a key part of the identity narratives of these anti-tourists. Munt (1994a) has noted the ways in which certain 'other' tourists engage in an 'alternative' style of travel, often claimed to be more 'authentic', in order to enhance their cultural capital and thus differentiate themselves, in Bourdieu's (1984) sense, from class fractions below and above them.

Indeed, the traveller/tourist distinction has run somewhat hand in hand, in academic discussion at least, with the debate concerning authenticity, and the idea that while certain tourists seek some notion of authenticity, others are quite content with contrived experiences.[3] Setting up this rather over-exaggerated opposition between the two, however, might prevent us from recognising, on the one hand, the contradictions and difficulties in tourists' attempts to form and maintain an appropriate identity and, on the other hand, tourists' ability to play with an appropriate identity. There is not necessarily a straightforward set of rules to follow, and the negotiation of an appropriate identity therefore presents a maze of overlapping and frequently contradictory musts and must-nots, which themselves are likely to be ever-changing. We can see this throughout the tourist portraits outlined above and also by looking again at the quote from the Belgian tourist shown at the beginning of this chapter:

> We are tourists but not typical ones. We try not to be, but we are. We try to go where there isn't so much tourism...The real tourists are going on buses, stopping, looking, drinking Coca Cola – whereas we spent two hours climbing in Zelve – so we're not really like tourists. We try not to be typical tourists, but still, we are...We are on the outside. Turks here live on tourism, so they look at us as tourists, to get our money...

While the central and overriding theme in the discourse of the tourists in Göreme is the expression of their keen desire to differentiate themselves from those whom they see as 'tourists', it seems that their identity is constantly in a state of flux. The negotiation of a non-tourist identity is clearly the main 'plot' along which, though with some variation, all of their stories run. The central character in the plot is 'The Tourist' (from now referred to in this chapter as 'Tourist' with a capital 'T' when intended to mean the category of Tourist that the 'non-tourists' in Göreme oppose themselves to), and the Tourist is an evil character with whom most of these tourists are keen not to be confused. It takes effort to achieve this, however, and they find themselves constantly walking on a knife-edge, because they always have the horrible suspicion that they are in fact indistinguishable from that character. Moreover, just when they think they might be doing fine – by coming to Göreme – they find themselves caught by the very people with whom they were trying to sympathise – the local people of Göreme – who most pointedly of all see them as Tourists.

So who is the terrible Tourist that these anti-tourists so keenly don't want to be? Dan from the USA saw the archetypal Tourist as someone who stays in three-star hotels, expects high levels of service and does not have proper inter-action with the local community. Max, the German backpacker, gave a similar picture, although he added that Tourists are always on tour, doing nothing but sightseeing, and they would want to eat German food all the time. An Australian woman interviewed provided a similar image: 'Tourists are the kind of people that go to the coast – you know, the kind of people who have no respect.' She continued:

> Well, it's a lack of education really. I don't like to say it but all those working-class English and Irish people, the kind of people who want bacon and eggs for breakfast. They shouldn't have to start cooking bacon here for people, not unless they want to of course, but they shouldn't have to cook bacon just to suit people who come here. I'm pleased that it has been quite controlled here so far – it's not really touristy yet.

Tourists are considered by these would-be non-tourists to be the hordes who go mostly to the sunshine coasts to engage in non-intellectual activities. They are generally older than non-tourists, and they have more money to spend on luxury and Western-style services while on holiday. If they do go to regions like Cappadocia, they are bussed around in large tour groups, following packed itineraries, and led by a guide who organises everything for them. They only go to the main museum sites, and there they 'paw at the frescoes in the churches', as one anti-tourist exclaimed, rather than having any real (intellectual) appreciation of the historical and cultural treasures they visit. They have no respect for the places and peoples they visit, and are not interested in any form of close or real interaction with those places and peoples. Finally, they have something of a bull-dozer effect on environments and cultures, through which they mindlessly plough, destroying everything in their path. All in all, it seems that the Tourist is a rather vile character and it is little wonder that the anti-tourist veins run so deep.

This oppositional theme running through the stories told by these non-tourists clearly reflects what Mowforth and Munt (1998) have termed the development of a *new tourist class*. New tourists, they say, who include the back-packers, the eco-tourists, the trekkers and the truckers, are engaging in this type of 'individual' travel almost exclusively as a strategy to differentiate themselves from class fractions below and above them. Termed 'ego-tourists' by Munt (1994b), they are searching for a style of travel reflective of an alternative lifestyle and one that enhances their cultural capital. Driven by a new culture of individualism, these tourists characteristically:

deem themselves unclassifiable, 'excluded', 'dropped out' or perhaps, in popular tourism discourse, 'alternative'; anything other than categorisation and assignment to a class.

(Mowforth and Munt 1998: 134)

They are, in other words, something of a counter-culture, and that is why they also consider themselves exempt from the criticisms often aimed at 'mass' tourism concerning so-called 'social and environmental impacts' caused in the tourist destinations. A large part of their narratives focuses on convincing themselves and others that, because of their respect for local cultures and environments, and because of their travel style being low on 'luxury' demands, they are not as harmful as 'evil' Tourists are to the places and peoples they visit.

This apparent concern for the 'other' and for not 'destroying the culture' of the people they visit, however, is closely intertwined with a concern for the tourists' own experience and their own identity. Note Akira's suggestion, for example, that Japanese package group tourists are remaining within the organised Japanese system, while, by travelling alone, he was escaping that system. A similar idea came from a middle-aged American woman who saw the 'tour bus' type of Tourist as someone who, even when on holiday, is 'moving at the pace of twentieth-century life, you know, quickly going from one place to another'. Dan's unease was very much focused on himself: he saw the Tourist as the victim of the locals' money-grabbing clutches, and the traveller as the hero whose adventuresome wit allows him to escape this terrible put-down. He found it easier to be this hero in Göreme than in more 'touristy' Ürgüp, although he wasn't entirely happy with the obviousness of Göreme's services for tourists either.

The trouble is, it seems, the more (non-)tourists there are in places such as Göreme, all trying to individuate their experiences and assert a non-tourist identity, the more they have to find new ways and new narratives to further differentiate themselves from the 'masses'. Mowforth and Munt describe this problem:

Of course travel has always been an expression of taste and a way of establishing class status. But, with the rapid growth in the numbers of people taking holidays, it has never been so widely used as at present. Put simply, the democratisation of tourism has created a social headache when it comes to classes attempting to differentiate themselves from one another.

(Mowforth and Munt 1998: 136)

One of the ways in which these non-tourists cope with the inevitable 'tourist' situation they find themselves in concerns their use of irony and their constant play with the category of the Tourist. For some it seems that the easiest way to cope is to admit that they are Tourists, so that they cannot be caught out when they fail to be otherwise: 'The real tourists are going on buses, stopping, looking, drinking Coca Cola...so we're not really like tourists...but still, we are…we've always had a European toilet!' Others twist the negotiation of a Tourist identity around into a sort of double-irony whereby they mock themselves for being Tourists, but also for attempting to be non-tourists. An example of this came in a young British tourist's comment that:

> it's hard to be a traveller in Turkey, isn't it? Because that's all about adventure and the cutting edge [he grinned], but here there are nice buses and plenty of *pansiyons*.

The speed at which trends are followed and dropped is rapid, and it seems now that tourists in Göreme are just as likely to mockingly resist a non-tourist identity as they are to mock a Tourist one. Elsrud also notes the adoption of an ironic tone, particularly with the women backpackers interviewed, as they talked about risk and adventure. Elsrud comments that 'irony, as a reaction to a collision between different systems of logic, has the intrinsic "advantage" of mediating between discrepant ideas' (Elsrud 2001: 614).

A central paradox written in to the ill-fated plot that non-tourists try to weave is that, although anti-tourists engage in a discourse that stresses 'uniqueness' and 'individuality', they require and actively seek the company (and audience) of other travellers or non-tourists for the support of their 'alternative' identity. If we recall, the reason that Julie and Sarah gave for getting off the bus in Göreme was that 'all the other backpackers got off the bus there'. Tourists like Göreme *because* it is a 'backpacker place', and an important aspect of their travel is about meeting other travellers because they are like-minded people. Some tourists visiting the village in the quiet season show disappointment that there aren't 'more people around' for them to meet – the only important people, of course, being other backpackers.

Similarly, most of the tourists do not enjoy staying in a *pansiyon* that is too quiet, and often follow word-of-mouth recommendations around the Turkey circuit concerning which *pansiyons* in Göreme are 'happening'. Consequently, as well as *pansiyons* tending to gather particular nationality groupings because of guidebook recommendations (as I mentioned earlier), they also attract particular 'types' of backpacker. Walking into some of Göreme's *pansiyons* in the early evening, one is confronted by crowds of people with trendy haircuts, lying around on the terraces drinking beer and playing cards or backgammon. Such *pansiyons* often have a foreign/tourist manager, whose knowledge of what these tourists want, such as laid-back ambience, plenty of cheap beer and a comfortable communal area in which to meet with other travellers, is often the secret to the *pansiyon*'s success. In other *pansiyons*, a quieter and more sober scene is taking place, with tourists reading, swapping tales from their day's adventures, and engaging in a political discussion, or at least a game of backgammon, with the Turkish owner of the establishment. In still other *pansiyons*, you might not find any guests at all because the current trends, with the help of the leading guidebooks, may have rendered those places not 'in'.

So it is that these tourists are all following the same routes and staying in the same villages and *pansiyons* along the way. Their 'out-of-the-way places' are only out of the way in as much as they are away from the package tour Tourist. Göreme, which is cited in travel guides and backpacker discourse as being a 'backpacker place', has become very much part of the 'scene' for these non-tourists, as it allows them to spatially and pointedly separate themselves from the Tourist, but at the same time to be with each other. They must seek out other non-tourists, partly to confirm that they are in the right places and doing the right kinds of things, but also because other non-tourists are the significant others with whom they share the common manuscripts used to negotiate and present their non-tourist identity.

This self-presentation often takes the form of a friendly competitive banter with each other regarding adventures had, places discovered and hardships overcome. In any case, for the large part, neither the folks back home at the end of the trip nor the Turkish people they are likely to encounter on their way tend to be very interested in stories of discovering a particular cave or how cheap they managed to bargain the museum entrance price down to. It is mainly the other like-minded non-tourists, then, who are the main combatants to engage with in this kind of competitive banter. As Ochs and Capps (1996) have noted, the audience is as much the author of a narrative as is the teller.

Much of the conversation among tourists sitting in *pansiyons* and restaurants in Göreme revolves around who has made the best 'discovery' by finding a valley or cave-church that no one else had found, or who got invited to a villager's cave home for a real cave-cooked dinner. The adventure narrative is highly valued among these non-tourists and this value very much depends, as Elsrud (2001) has pointed out, on the mythologies that surround them. The idea, for example, that a dinner invitation to a Göreme cave home is a more authentic experience than a dinner invitation to a family home in Milton Keynes is a mythical idea. A crucial point to add here, however, is that these experiences, such as a dinner invitation to a Göreme cave home, need to actually *happen*, and they tend to work best both in the experience and in the telling if they happen by surprise.

Of course, such surprises should not be too unpalatable or downright dangerous as to be unmanageable: the public bus that they travelled to Göreme in should have been *on the verge* of going over the cliff-edge, rather than actually going over it. And most non-tourists would get bored and thus find it rather unpalatable if invited to stay for a week with a Turkish family in a non-touristic village that really was off the beaten track. What is more, they cannot spend too long away from the company of other backpackers, because they need the regular presence of an audience to whom they can narrate their experiences. So it is that by and large they stick to the backpacker route and stay in backpacker places. However, while they are travelling on that fairly main road they require a certain level of surprise events, or happenstance, in order to individuate their particular experience on that road. They therefore have to travel in such as a way as to be open to unplanned and unpredictable happenings, open to possibilities. There are various ways that they do this: the avoidance of the pre-planned tour; doing things as cheaply as possible; travelling in a particular temporal style; and going to and staying in particular places. I will now discuss each of these in turn.

Avoidance of the pre-planned package tour

> I don't take tours because I prefer to be free to being guided and controlled – then I can explore, and climb up tunnels whenever I want to.

> We're not 'tour people', it is better to discover it for yourself. It's like exploring as a child, you feel as though you will find something that no one else has found.

These extracts from interviews with tourists in Göreme, together with many elements in the portraits outlined earlier, show that an important theme in non-tourists' oppositional relationship with organised Tourists is centred around the issue of the 'tour'. As I pointed out in the latter part of the previous chapter, the travel style of the tourists in Göreme is opposed to what Bruner has termed the 'touristic surrender' of the tour group whereby the tourists must succumb to a fixed itinerary and to the control of a guide. One afternoon at the Göreme Open-Air Museum, I listened to two backpackers as they watched and commented on the hordes of tourists following their guides. The Australian traveller remarked: 'I don't like being told what to do. At home I have to work, so on holiday I want to be free to do what I want, when I want.' The other, who was from Singapore, contested, 'But they [package tourists] are very relaxed. They don't have to make any decisions. They just follow, it's very easy.' The Australian replied,

> Yeah, but that's exactly what I don't want to do! To follow other tourists around all day, and to have to listen to all that boring information is not my idea of a good time.

Despite their general dislike for this method of sightseeing, however, many of Göreme's tourists do take tours of the Cappadocia region, indicating another contradiction in their quests and identity. There are somewhere between fifteen and twenty tour agencies operating in Göreme village, all selling day tours to non-Tourists. Those who only have a few days to spend in Cappadocia find it far easier to visit the main historic sites, such as the 'underground cities', by organised tour. And because images and myths in the travel literature tell tourists that they must see certain sites, as discussed in the previous chapter, a tour often becomes something of a necessity to ensure that nothing important is missed.

This point further highlights the plurality in these tourists' quests and experiences, and also that it is because these different quests and experiences may clash and contradict each other that these tourists' non-tourist identities must constantly be negotiated and re-worked throughout their travel experience. Clashing with their urge to see the main tourist sites during their trip, non-tourists are driven by the 'cult of individualism', as Munt (1994b) has it, and a desire to have what Urry (1990) refers to as a 'semi-spiritual relationship' with the tourist object. Urry calls this the 'Romantic gaze' of the tourist, as opposed to the 'collective gaze' which is that of the hordes who travel on large coach trips. An illustration comes from an interview with a lawyer from the USA who the previous day had conceded to taking a day tour of the main sites around Cappadocia, and afterwards compared that to wandering around freely:

> Wandering in the back streets of this town, and the farms and the fields, you get a feeling that there's a lot more to it, you get like a slice of what it really is here, as opposed to yesterday on the tour, or walking through the open-air museum where you're surrounded by tourists, all pawing at the frescoes, and here in the standard route, with the churches and all that stuff, and 'here's the next stop and you can buy this and that'. So, having the chance to see the normal stuff is what I really wanted to get out of coming here. I really don't like tours, any of them.

Similarly, a Canadian told me, while we were hiking in a Göreme valley:

> In the valleys I feel as though I'm discovering something new. In this valley, and with this view, I feel as though it's all mine! Whereas, in that underground city, I knew that there were thousands of other tourists going through.

The individual and unique experiences the tourists in Göreme desire in order to individuate their travel experiences by definition cannot be had on a 'group tour'. In addition, they want to be able to frame their experience for themselves rather than having it staged and pre-framed for them. This is explained by the Canadian in a conversation about why he preferred wandering freely to being on tour:

> When I was on the tour, I was told what I was looking at – 'this is a panoramic view'. I was told what to do, I was guided through the whole experience, whereas part of seeing it on your own is the idea of creating your own internal text of what the world's made up of. In a tour, if I ask 'what do I see?', the answer is given to me. Out here in the valley the answer isn't so given – I have to get it, I have to create my own idea of what's going on. I get a sense of mastery from this, expanding myself with new experiences.

If the experience or encounter is already decided on and packaged as a touristic event, then these tourists lose their own sense of framing and subjectivity and

Plate 3.1 Exploring in the valleys around Cappadocia.

their interest is lost. Pinney (1994) makes a similar point in his discussion of 'virtual travel' created through computerised simulators, arguing that the ability to frame one's own experiences is critical to the consolidation of the self. The key problem with virtual travel is that there is nothing that can happen which has not been pre-written into the programme. Or, as Ritzer and Liska put it, 'there are no surprises on a virtual tour' (Ritzer and Liska 1997: 101). The pre-planning on a package tour also, though to a lesser degree than a computer-simulated tour, limits the possibility that the unplanned may happen; all the tourists know that the day's events have been undertaken by the guide numerous times before and that nothing, or very little, is new or left to chance.

Moreover, tourists enjoy being able to clamber and explore in Göreme's valleys, caves and tunnels, and the area is described by tourists as being 'like a huge adventure playground for adults'. So, while the Göreme landscape is on the one hand a spectacle to be gazed upon, it is also a place to get into, to *interact* with. The ascendancy of the eye over the other senses has led to an over-visualisation in tourism and the idea of the tourist gaze.[4] According to Little, tourist discourse, set up by tour operators and tourist entrepreneurs, fashions itself as a mass-mediated visualisation…Tourist productions of all sorts 'focus' on what the tourist sees.

(Little 1991: 149)

Indeed, the representations of Cappadocia discussed in the previous chapter demonstrate this 'mass-mediated visualisation', highlighting also the pronounced visualising tendency in the 'primary tourist gaze'. This is a large part of the problem with that primary gaze. With grazed hands and bruised knees, Göreme's tourists coming back after a day of hiking and clambering through caves and fairy chimneys certainly looked vibrant in comparison to those climbing out of a mini-bus after a day's tour of the prescribed 'sights'.

Most of all, though, the avoidance of pre-planned tours, and the primary tourist gaze they perpetuate, opens the non-tourists in Göreme to the possibility of encountering the unexpected; of coming across a hidden cave-church that was not written about in the guidebooks, or of being invited by a farmer to eat some fruit off his trees. They enjoy allowing, even inviting, the object of their interest – the locality and the local people – to be active in the writing of their travel experience. An example of this was described by a tourist from New Zealand who had made the effort to rent a motorbike in order to visit the 'sights' around the Cappadocia region rather than taking a daily mini-bus tour. He said that being on the bike he had enjoyed the chance to stop in villages 'where there was all sorts of life going on'. He continued:

> I was invited to a wedding in one village and to join in Turkish music and dancing in another. Little boys crowding round the bikes and asking for rides. It was much better than a fixed tour.

Many tourists who avoid tours and other 'packaged' experiences they associate with the Tourist shroud this avoidance in a discourse concerning cost and budget. The narratives of the tourists in Göreme are full of comparisons of the prices for things. Riley has also noted among backpackers this 'all-encompassing, at times almost obsessive, focus on budgets' (Riley 1988: 320), observing that status among travellers is closely tied to living cheaply and obtaining the best 'bargains', which serve as indicators that one is an experienced traveller (ibid.). What obtaining bargains and travelling as cheaply as possible might also do for these non-tourists, though, is increase the level of unplanned and unpredictable happenings.

Budgeting for possibility

According to the non-tourists in Göreme, Tourists pay high prices for the comfort of large Western hotels and specially chartered air-conditioned coaches, Western meals – they 'just want to eat German food all the time' – and the safety of having a guide by their side who understands their culture as much as that of the place they are visiting. Non-tourists, in contrast, take the cheaper options of local public transport, more basic 'backpacker' accommodation and food from local restaurants. It grates terribly with them if they are treated by tourist shops and restaurants as 'a walking dollar', because that is to be treated as a Tourist and they definitely do not want that. The more they can escape this terrible put-down by avoiding paying Tourist prices for goods and services on their travels, the better.

The competitive banter that the non-tourists in Göreme engage in often involves comparing how much was paid for this or that and even, sometimes, how much was got for free. One day I overheard a woman from the USA sitting on the terrace of her *pansiyon* telling other tourists how she likes Turkey because it is cheap. She went on to explain how, whenever she arrives in a new place, she 'begins to flirt with some guy' so that she will get free meals and accommodation for the rest of her stay. Although this woman is a rather strong case in point, there is something in her quest for 'free' services that rings true for most of the tourists who stay in Göreme.

This reluctance to spend money has important implications not only regarding these tourists' relationships with the Göreme people but also regarding the Göreme entrepreneurs' relationships with each other. Those implications are discussed in Chapters 5 and 6. What is important in terms of understanding the tourists' behaviours and narratives is the way in which this quest for cheap and free services opens them to adventure and happenstance. Again, many of the adventuresome connotations attached to public transport and local restaurants are little more than mythical ideas: there is no real reason why the public bus is more likely to (almost) go over the cliff than a tour coach and, by all accounts, there is probably less chance of contracting food poisoning from a busy local restaurant than from a hotel dinner buffet that has been sitting there warm for a few hours. What really is more likely to happen on a journey by public bus, however, is a chance interaction with a local family who invite you to come home and eat with them, or that you miss the evening's last connecting bus to your intended destination and so you end up staying in a town you had previously never heard of, and having an interesting time and an even better story to tell afterwards.

The reasoning behind the budgeting of the 'budget traveller' is not necessarily because they are any poorer than their loathed counterparts, the Tourists, but perhaps it is rather because of the opening up to uncertainty, and indeed serendipity, that budgeting can achieve. For the most part, the less paid for something the less certainty goes with it. While 'roughing it' for a dollar on the roof of a *pansiyon* rather than paying for a room, you might get soaked by a sudden thunderstorm, but at the same time you might see the most spectacular display of lightning ever. And although hitchhiking from Göreme to the south coast might be slower than taking the bus, it will undeniably bring more surprises.

Temporal opening to possibility

It is clear from the portraits above that the tourists in Göreme contrast their own temporal behaviour to that of package group Tourists, who they see as being on short trips and therefore whizzing through Cappadocia in a couple of days following a full and strictly organised itinerary. The non-tourists in Göreme generally prefer to travel for longer periods and with a much more relaxed schedule, or preferably no schedule at all. They spend much of their time in the village 'hanging around' in their *pansiyon*, preferring not to have to do anything at any particular time. In their travel narratives they frequently use terms such as 'wandering', 'exploring' and 'hanging around', and some make a point of not wearing a watch while travelling, saying they prefer to sleep and eat just when they feel like it. A lawyer from the USA travelling in Göreme told me:

> I live a highly scheduled life, so I will do things that, you know, like not wearing a watch for days, and I've literally gone and taken a ferry and I had no clue where it was going, or the next bus. It's the absolute polar opposite, and a sort of balancing.

Akira, too, commented that he had come travelling in this style because he wanted to be released from the 'very systematic country' that he comes from.

The notion of tourism as a release from the temporal structure of modern life has been discussed by Cohen and Taylor (1992) and Wang (2000). According to Wang, it is the 'routinization' of time, and also its segmentation – such as between work and leisure time – that has led, under the conditions of modernity, to a sense of temporal alienation in everyday life:

> Routinized work in industry or a bureaucracy imposes a constraining, compelling, and rigid tempo and rhythm, a situation in which individuals become automated, robot-like, de-individualised, repetitively doing Sisyphus-like wearing tasks. Toiling under such a working rhythm, employees' acts, pace, and speed are set by machines and manager's scientific calculations...Under such conditions workers experience temporal alienation.
>
> (Wang 2000: 105)

This temporal alienation manifests itself as feelings of boredom and monotony according to Cohen and Taylor: We become disturbed by 'the predictability of the journey' and 'the knowledge that today's route will be much like yesterday's' (Cohen and Taylor 1992: 46). Tourism, then, can be seen as a response to and as a temporary escape by tourists from the routine, the mundanity and the boredom of their everyday lives.

Certainly, the tourists in Göreme try to avoid any sense of temporal organisation on their trip, leaving themselves open to following whatever road may present itself to them. Many backpackers spend long periods of time travelling so that, if the right circumstances arise, they are able to stop and 'hang out' in a place for a few days or weeks or even months. Some tourists I interviewed described this as staying in a place long enough that they felt themselves moving out of the realm of Tourist experience. The following is an example from an Australian woman who stayed in Göreme for a few months working in restaurants:

> I could have just passed through Göreme in five days and just thought Göreme to be the peaceful, quiet village. You know, I wouldn't have got to know some of the Turkish people, wouldn't have understood their religion or their culture. I'd have just passed through and said 'Yeah, I've been to Turkey'. But now I feel, I mean I've lived here for seven weeks. I know and I've understood – well tried to understand – you know, the Turkish culture and that. So it's really been a good experience for me, a cultural experience.

Similarly, a Canadian tourist who stayed in the village for a month noted that through time he began 'to engage with Göreme more as a living village'. Staying in one place for a while is seen by these non-tourists as allowing them to achieve a fuller and more 'real' understanding of the villagers and their 'culture'. Similar to their desire to go beyond the visual gaze and have a more exploratory, and even playful, interaction with the physical landscape of cave- and tunnel-filled valleys, therefore, is their quest for a fuller level of interaction with the local villagers in Göreme.

Interacting with the authentically social

> I wanted to go to a place, I think like lots of tourists, that lots of tourists didn't go to – that has people living in their natural habitat – so that I would actually be able to see the culture, as opposed to having to be with thousands of tourists...So my view of coming to Cappadocia was that, you know, go to Istanbul and see the sights, and I had a view of that as being very touristy and city-like and this would be more, kind of, really seeing what the culture was like in a very different kind of way. So I was more interested in seeing, you know, living people, seeing what it was like – as much as you could when you're travelling through. Also, when we got here, it seemed like a much calmer place, quieter, compared to being in the city.
>
> (Female tourist, a psychologist from the USA)

There are two slightly different but overlapping strands in this woman's explanation of why she wanted to visit Cappadocia. On one hand, the woman wanted to see 'people living in their natural habitat'. On the other hand, she wanted to 'actually be able to see the culture, as opposed to being with thousands of tourists'. The first refers to something that is external to her, a quality inherent in the place and people she is visiting, and is based on what Selwyn (1996) has termed the myth of the 'authentically social'. The second is more about the quality of the tourist experience and the nature and quality of the tourist's interaction with the perceived 'authentically social'.

Both of these ideas are clearly present in most of the portraits of tourists outlined above. In Alison and Clare's contrasting of Göreme life with the 'rat-race' of life at home, for example, they see both their own life in Göreme as relaxed and unscheduled relative to life back in Sydney, and the life of Göreme village more generally as contrasting with the rat-race of modern, and perhaps urban, places. Akira, too, perceived the Turkish lifestyle as having 'lots of time' compared to Japan where the 'system is now complete' and 'time is money'. A similar idea is highlighted in the following extract from an interview with a Belgian tourist in which she explained what she liked about being in Göreme village:

> It is very beautiful. It's amazing that they live in houses in the rock still at this time, in days when we have electricity and so on. It seems so cosy, it's quiet, it's all very at ease, there's no stress. Also the Turks' way of life is completely different. They are all at ease, they can take their time – they don't seem to have any worries – you can imagine that nothing needs to be done, you don't need to be on time. It's all so cosy. Like in Zelve where they all have holes in the rock, so that everything is open and they have a lot of contact with each other. We have our fences, here they are more of a community, living on top of each other – literally! So it's the contradiction with what we have at home. It's really nice to take your time – you don't feel obliged to buy, there's give and take. Even when you see obvious competition between the businessmen, and then you see them sitting together and drinking tea – you can just imagine them like that in their caves. It is so different from home.

So, although the people of Göreme have been experiencing social change and interaction with outsiders for a long time, particularly through outward migration to urban centres in Turkey and northern Europe,[5] tourists imagine Göreme to be stuck in a pre-modern time. Furthermore, they hope that they have caught it just in time before those elements of pre-modern life disappear. This was illustrated to me when a German couple who were visiting Göreme for the second time expressed sadness at the observation that the village had become 'modernised' in the space of just one year. The previous year they had seen many women out cooking on open fires in the streets and courtyards, whereas this year they had not seen even one example of this 'traditional' activity. The conclusion drawn from this observation was that over the past year the villagers had earned enough from tourism to fit modern kitchens inside their houses and so this Romantic tradition had, from one year to the next, been lost. In actual fact, the tourists had been in Göreme a month later in the season the previous year and had chanced on the time in the autumn when villagers use the last crops of grapes to make *pekmez* (grape syrup). At *pekmez*-making time village women are indeed out in open spaces in the winding residential streets boiling down the syrup in huge cauldrons balanced over large open fires. The second year these tourists came they were too early to see the *pekmez*-boiling and they concluded that it was a lost tradition: hence the 'trope of the vanishing primitive' (Bruner 1991: 243).

These ideas resonate strongly with the tourist 'cult of the primitive', discussed in Chapter 2, and with the Romantic tourist quest to experience the 'other' who still lives a simple, authentically social life that is free from the stresses and woes of modern life:

> The character of this other derives from belonging to an imagined world which is variously pre-modern, pre-commoditized or part of a benign whole recaptured in the mind of a tourist. This is a world which is eminently and authentically social. Thus what makes a tourist destination attractive is that it is thought to have a special 'spirit of place', which derives from the sociability of its residents. Or, to put it another way, in successful tourist destinations the natives are friendly.
>
> (Selwyn 1996: 21)

These ideas could not be more closely backed up than they are in the following words from a middle-aged lawyer from the USA who has been coming back repeatedly to Göreme for a number of years:

Plate 3.2 Boiling *pekmez* (grape syrup).

I feel really at home here, there's something magical about this place. Part of it is the incredible way that this community deals with the indoor and the outdoor. There's such a natural relation with the environment, and I don't just mean because it's all in caves, but there's a kind of easy communion between the indoor and the outdoor in the way that the people live, and the way the architecture is done and so on. And I found the people to be incredibly friendly the first time I came here.

There is the idea here that the pre-modern authentically social lifestyle perceived by tourists to be lived in Göreme will somehow rub off on them, easing them from the tensions, stresses and monotony of their 'rat-race' life at home. Such feelings and tourist quests have been associated with a wider climate of what has been termed a 'post-modern' nostalgic yearning (see, for example, Jameson 1991 and Urry 1990), and they are very much implied in MacCannell's idea that tourists attempt to 'overcome the discontinuity of modernity' (MacCannell 1976: 13) by experiencing authenticity in other (pre-modern) times and other places. Viewed in this way, the travel experience becomes akin to a kind of therapy, as is expressed in the continuation of the interview with the American lawyer quoted above:

I find it hugely comfortable, sort of walking around, grinding back into natural rhythms, and for me it helps enormously, and that's the whole reason why I take these vacation trips, 'cause I get back and I look out of my high-rise, and I see people screaming all around, and I just know it's not that important, you know. People do just fine on 99 percent of the earth's surface and it's just not that important...I find that I have a perspective on my life when I step back from it and travel, and it's so interesting seeing other people's perspectives on things – it's absolutely fascinating. Each time I come back, I think I'm a broader and wiser and better person, and a much calmer person, every time I travel. In fact, it's interesting that people now remark that I've become much calmer. Travel's been a big part of my maturation process.

Moreover, it is therapeutic to meet with other like-minded travellers and to feel a sense of group or community together with them, as so many of the tourists in Göreme do. A comment left by a tourist in a Göreme *pansiyon* visitors' book summarises this point:

I seem to have stumbled on an Australian expatriate community – maybe some sort of homing instinct brought me here. Cappadocia is perhaps one of the highlights of Kathmandu to Istanbul (yes, I am a neo-hippy!). Restful, sleep in my cave, a real re-birthing experience. Friendly people, no hassles. Thanks – Campbell, Australia.

The use of the term 'stumbled on' again suggests the notion of serendipity, highlighting the importance of the serendipitous experience of meeting other travellers along the way, and also of gaining a sense of *communitas* from spending time with them.[6] This is further illustrated in the following extract from my field-notes on an observation I made concerning the tourists that stayed in a *pansiyon* in the village that I frequently visited:

> All three of the guys that I've met there this week, who coincidentally are all sharing the same dormitory room, are travelling to get away and to sort things out; to be alone but also to speak to other travellers who are unconnected to their home situations. They've all been staying here for a week to ten days now, hanging around the *pansiyon* and going for long walks together in the valleys. They have no interest in the village, as such, nor with interacting with Turkish people. They've all become busy sorting out their own and each other's crises.

For some non-tourists, it seems, the experience of the 'authentically social' and the importance of happenstance lie much more in who they meet along the way and the 'community' within *pansiyons* and tourist bars than with ideas and images concerning the pre-modern lives of the local people. For most, though, it is more likely a combination of both, whereby the local people and place provide an enhancing backdrop for a sense of *communitas* to be had from meeting like-minded travellers, but they also embrace chance meetings and happenings that are particular to Göreme and the Göreme people. Hanging around in a place where time itself seems to stand still, and where the tourists themselves have nothing particular to do nor anywhere in particular to go opens up the possibility of happenstance. It gives tourists more chance of hitting on a wedding in the village or the procession for a boy's circumcision ceremony. It allows them to follow through with people they meet and to take up serendipitous invitations they might receive. It allows them to explore in the Göreme valleys and to see where a path might lead them. All of these chance happenings in turn serve to individuate their experience and memory of Göreme, and contribute to the experience and identity narratives of these non-tourists.

The search for serendipity, not touristic surrender

By considering the important part that interaction and happenstance play in the travel narratives and identity negotiations of the tourists in Göreme, it has become possible to understand the 'second gaze' of these tourists. This gaze is an interactive one, and therefore a negotiable one, as it relies very much on the tourist having chance encounters. By no means is this second gaze completely unscripted, however, and by no means does it cancel out the first, primary gaze. The motivations of these tourists in their coming to Cappadocia are undoubtedly grounded in the images and myths discussed in the previous chapter concerning the 'Romantic' features of the region. They want to experience the 'extraordinary' landscape and culture just as other tourists do. As one tourist said, 'Göreme is one of the places you have to go to when you're in Turkey.'

In this sense, these tourists are most certainly following the (back)pack, and guidebooks such as *Lonely Planet* and *Rough Guide* play an important role in making sure that their readers know what and where the 'in' scene is. Concurrently, however, these non-tourists are engaging in a post-Fordist resistance to 'mass' Tourist identity and gaze through resisting a mass structuring and organising of their experience. Rather than surrendering to the primary gaze neatly arranged and represented by the package tour and associated 'industries', they attempt to move beyond that gaze. This attempt thus becomes their new script, or their alternative script, and that script is to travel in ways that will open their journey up to possibility – to serendipity – in interactions with the places and people they meet. Those serendipitous events then become incorporated into their travel narratives and serve to individualise their experience and identity.

A question might be asked here regarding the place of serendipity in the experiences of repeat visitors to Göreme. Certainly there are tourists, such as Bill the US lawyer, who return again and again, and their experiences of and in Göreme will surely become more predictable each time they return. However, people like Bill continue to travel following the particular travel 'styles' and scripts that open them to happenstance. The experiences had and the adventures encountered in each visit would therefore always be very different from the last because of their contingency on current and unexpected happenings in the village, as well as on the people they happen to meet each time, whether tourists staying in their *pansiyon* or villagers. Furthermore, there seem to always be more churches to discover and more valleys to get lost in. Cappadocia is an adventuresome place indeed.

These non-tourists do not want their experiences in places they visit to be completely predictable and controllable. They want those places to speak back to them, to surprise them, to challenge them, in order to satisfy their quests for serendipity and experiences that are suitable for narration. Their actual experiences, and identity, therefore, are not fully prescribed but, to the contrary, are negotiated and constructed through their interactions with the places and peoples they meet. The negotiation and interaction between the tourists and the Göreme people and place is discussed further in the later chapters of this book. In the next two chapters I will turn the focus onto the people of Göreme and the discourses and practices they bring to their interactions with tourists.

Notes

1   MacCannell has recently identified this other gaze, referring to it as the 'second gaze'. In his discussion of the second gaze, he has attempted to regenerate a notion of 'tourist agency', unlocking the tourist subject from the 'prison house of tourism' (MacCannell 2001: 24).
2   For example, Buzard (1993), Dann (1996, 1999), Elsrud (2001), Macleod (1997), Munt (1994a), Riley (1988) and Tucker (1997, 2001).

3   See Baudrillard (1983), Boorstin (1961), Bruner (1989), Cohen (1979, 1988b, 1995), Lash and Urry (1994), MacCannell (1976), Pretes (1995), Urry (1990) and Wang (2000).

4   For critiques of this over-visualising tendency in tourism theory, see Adler (1989), Game (1991), Little (1991), Tucker (1997) and Veijola and Jokinen (1994).

5   Outward migration and social change are mentioned in Chapters 1 and 4. See also Abadan-Unat (1993), Kandiyoti (1991), Magnarella (1974) and Schiffauer (1993).

6   It has been noted by Graburn (1983, 1989) that tourism bears a resemblance to ritual in that its role is to break up the profanity of modern existence. An aspect of ritual, as described by Turner (1969), is the *communitas* – the experience of an intense togetherness with other ritual doers.

4 Continuity and change
Gender and production in Göreme

Yazın, türizm.
Kışın, kuru üzüm.
(In summer, tourism. In winter, dried grapes.)

Tourism now plays a prominent part in Göreme life, most obviously throughout the summer. The rhyme above, which is often repeated by the Göreme people, thus summarises the way they see their lives today. The rhyme is expressive of a duality concerning the villagers' economic and social activities; not only a seasonal duality, but also one based on gender segregation. As is reported in other Turkish villages, the household is the central unit of social organisation in Göreme and, in accordance with Islamic codes and practices, there is a strict segregation of the sexes with a well-defined distribution of economic and social activity according to gender upheld by principles of honour and virtue.[1] Most households continue to own gardens or small fields that are worked, mostly by the women, with the use of mainly simple technology and through a variety of labour-exchange networks. Production in these gardens is mainly for home consumption, and any surplus is sold at the market, the profits going towards the purchase of basic household goods and the payment of taxes. Concurrently, Göreme today is inextricably linked to national and global processes, both social and economic. Through tourism and migration, the men of Göreme have become entrepreneurs and wage-labourers, even if their households continue their subsistence farming activities.

This chapter (together with the following one, which focuses more specifically on the tourism business in the village) describes the variety of Göreme lives. The aim here is to explain why it is that men are more directly involved with tourism business than women, and to develop a sense of both the continuity and the social change in the village. This chapter will thus serve to form a background understanding for later chapters regarding how and why the variety of villagers deal with and experience tourism, and their interactions with tourists in the ways they do. In the discussion I will at times refer more closely to particular members of the village and their immediate family groups. I begin by introducing those characters, and will use their cases as signposts for discussion as the chapter continues.

Göreme lives

'Anne'

'Anne' ('Mother'), as I knew her, is elderly and, having become completely
blind, has been looked after by her daughters and daughters-in-law since the
death of her husband some years ago. She lives in a small concrete house at
the lower end of the village, one of a few houses that were built by the
government in the 1960s in order to re-house those families whose rock-cut
houses were crumbling and considered dangerous. I came to know her family
well, and sometimes helped them in the fields, or sat chatting and joking with
them in the evenings. The younger women would often make fun of their
blind old mother, tricking her in her blindness. One evening in late summer
when the grapes were laid out to dry in the fields (the few days of the year
when villagers pray that they do not have rain), we sat outside in the garden
with 'Anne', and sprinkled water from a bowl over her and feigned panic in
the mock rain. 'Anne' seemed used to such jokes, though she frequently
complained that her blindness had now rendered her unable to participate
with her daughters in the gardening work and food-production activities of
the household. Those activities had formed the most important part of her life
and she often considers her life with an air of nostalgia, though she also
frequently remembers 'the hard times' when she worked in the gardens while
her husband tended sheep up on the mountains. In those days she had to
cook and sew by the light of an oil lamp within the dark cave of their old
Göreme house.

'Anne' has three sons and two daughters. All three sons work in tourism
in the village. The oldest son, now in his late forties, lives with his wife and
teenage children in the old family cave-house. He keeps horses in the caves
near his house and runs horse-riding trips through the Göreme valleys for
tourists. The youngest brother has owned and run one of the longest-
running *pansiyons* in the village, together with his Scottish wife, the first
'tourist bride' to marry into the village, since 1984. Among her many grand-
children, 'Anne' has two bilingual half-Scottish granddaughters.

The middle son of 'Anne' is Abbas, who runs a tour agency connected with
the *pansiyon* of his younger brother. As I said in the introduction chapter,
Abbas adopted me as his 'niece' while I was in Göreme, so I spent a lot of
time in the company of himself and his family. Abbas is in his mid-forties and
has lived in Göreme all his life. Before running the agency he owned and ran
restaurants in the village, and he prides himself on the fact that he opened the
first 'tourist restaurant' in Göreme in 1977. The tour agency he runs now sells
daily mini-bus tours around the sites of Cappadocia, hires out mopeds and
exchanges foreign currency. This is where Abbas spends much of his time
during the summer months, though, because he employs university students
to work throughout the summer season, he is fairly free to come and go from
the agency when he wants in order to visit friends doing similar work in other

tourism businesses or to go fishing at a river ten kilometres from Göreme. Whenever necessary he also helps his wife with the gardening work in their five gardens or orchards. During the extremely cold and often snowy winter, when few tourists pass through the village, there is not much work to be done in the agency. Nor is there anything to do in the fields, so Abbas sleeps long hours and spends most of the day chatting and drinking tea with friends in the tea house at the centre of the village.

Ali

Ali has, like the sons of 'Anne', worked in tourism all of his adult life. However, he is approximately fifteen years younger than Abbas and so tourism had already begun when he left school at the age of eleven. I came to know Ali well because he was working in two of the places where I lived for parts of my fieldwork time, and I learned a great deal from his conversations and behaviour. As a teenager Ali was a cook for an 'adventure' tour company and travelled around Turkey cooking for 'adventure tourists'. Then, through most of his twenties, he cooked for friends' *pansiyons*, until in 1997 he was invited to work as a crew member for a hot-air ballooning tourism business newly set up by a couple from northern Europe.

Four years ago, Ali married a tourist he met in the *pansiyon* where he was working at the time. His wife is South African, and being of part Malaysian descent is Muslim, something that Ali considers important. They have two young daughters and live in the old family cave-house with Ali's elderly mother. On his father's recent death, Ali inherited the house in order that he would eventually turn it into some sort of tourism business. All of the family land was passed on to and is worked by Ali's older brother, Mehmet, who has never been a tourism entrepreneur. For most of his working life, Mehmet was a civil servant, initially working for the police and later as a clerk in the court house in the nearby town of Avanos. These two brothers highlight the link to be made between generations of village men and levels of involvement with tourism. Mehmet explains the difference between his own career and his brother's as follows:

> Well, in our time, when I was a child, tourism was not as developed as it is today. People were poorer. It was not like today. We had to work in our gardens and fields. There wasn't any tourism. And then I went to the army for two years, and later I chose my official job. I preferred that. It was after those years that tourism developed...slowly. Tourism is still in the process of developing. As Ali was growing up, tourism was growing up too at the same time. Ali grew up with tourism.
>
> (Translation from the original Turkish – from now on such translations will be marked [*trans.*].)

I asked Mehmet if he had ever tried to live in Europe:

Yes, before I applied to a job agency in order to find a job in Europe, but I was told that they were by then over their capacity and I was rejected. So I couldn't go. And then there was a textile factory in Nevşehir, and I wanted to work there but I was told that it was difficult to get a job there so I gave up...The job in Avanos was my fate.

[*trans.*]

Unlike many of the Göreme adult men in the late 1960s, Mehmet seems to have been unlucky in his bid to migrate to Europe. As with many other parts of Turkey, Göreme at that time saw quite a large exodus of its working population. It is estimated that by 1970 as many as a hundred Göreme families had migrated to the northern European countries of Germany, Holland and Belgium.

Esin

I met Esin because I used to pass her house regularly and she would be sitting on the doorstep of her own or neighbouring houses, together with her friends, neighbours and their mothers. These women and girls would usually be busy crocheting or knitting, or preparing vegetables or fruit, but also sitting where they could observe passers-by, some of them tourists. Göreme women's lives are strictly governed by codes of shame and honour, and are very much centred around the home and the immediate neighbourhood. Esin attended primary school in the village, but did not go on to high school because that would have meant her travelling out of the village to Nevşehir every day. One of her brothers attends university in Izmir on the west coast of Turkey, something that would be unthinkable for Esin. Now in her late teens, Esin will probably be married within the next couple of years.

From spring to autumn Esin spends much of the day working with her mother in the family gardens in the valleys surrounding Göreme. Depending on the season, this work may involve tending and harvesting grapes, wheat or apricots, as well as gathering grass for the cows and donkey, and the gradual collection of wood-fuel for the winter. In the afternoon they come back home and, after doing any cleaning and food preparation needed that day, get together with friends and neighbouring women to sit and chat. Later on, the men return from the village centre for their evening meal: Esin's father and brother from their tourist souvenir shop, and her grandfather from the tea house and mosque. Most of the women pray five times a day in their homes, although these days some younger women, Esin included, have slackened in this regard.

In Göreme, midsummer is the time for marriage, as this is when many of the people from the village who had migrated to Europe return for a summer vacation. Esin attends many such parties, each lasting over three days, where groups of women gather inside or just outside the courtyard of the bride's family's house and dance with each other to a cassette playing Turkish folk music. Weddings are virtually the only formally organised opportunity for village women to have 'enjoyment' (*eğelenmek*), making them the most conspicuous social events for women of the village. There is usually much excitement among the unmarried girls concerning attendance at such parties and great care is taken that they look their best, for this is where they may be spotted by potential mother-in-laws.

As part of my fieldwork, I gave a small camera to Esin and asked her to record with it the main aspects of her life. This photographic activity was particularly appropriate to my fieldwork because of the part that photography plays in any case in touristic settings such as Göreme. I hoped to give Esin, in the face of being always represented by others (tourists and myself), a chance to represent herself through photographs. Of course, such a method was by no means conclusive in its attempt to capture Esin's own representations of her life: she is likely, for example, to have taken pictures that she thought I myself wanted and would have taken. She may have taken images that she construed to be generally most appropriate to 'photography'. However, it is possible to draw some modest conclusions from the general contents and frequency of certain types of images that she collected.

First, all of the photographs were taken either in or close to her home, and twelve out of twenty-five of the photos were of Esin's female friends and relatives. These reflect the spatial segregation that exists between the genders in Göreme society: the lives of girls like Esin are centred around the household, and women and girls operate almost entirely within a female milieu. Second, while a few of the photographs show the girls relaxing, a higher number depict work and food-production activities: Esin's mother kneading bread dough, a group of six women neighbours 'communally' baking bread, some neighbours chopping wood ready for the winter, and a pile of dried grapes stored in the cave cellar of Esin's house. These pictures convey the integral part that food production plays in the lives of women in the village. Lastly, nine of the photographs show panoramic views of Göreme, of roof-tops and 'fairy chimneys', all taken from near Esin's house, which is situated at a high point of the village. Although Esin may have been influenced in this by her observations of tourists taking pictures of panoramic views of Göreme, these pictures might indicate that, in depicting her life through the photographs, Esin was urged to depict Göreme the place. A sense of belonging to Göreme (being 'Göremeli') is central to villager identity.

Plate 4.1 Example of Esin's photographs: the view from the roof of Esin's house. A
woman lays out grapes to dry on another roof-top.

These short vignettes of Göreme characters depict the duality that exists
between the lives of women and the lives of men. While men spend their days
dealing with their tourism businesses, or other activities situated in the centre
of the village, the women have remained close to the household in their daily
activities. To refer back to the rhyme at the beginning of the chapter, the men
are largely engaged with *türizm*, while the women are engaged with *kuru
üzüm* (dried grapes). The gendered division of social and economic activities
as tourism develops in Göreme are thus central to an understanding of
tourism relations in this rural Turkish village.[2]

Also highlighted in the vignettes is the remarkable level of social and
economic change that has occurred in Göreme over the past decades. The
memories that Abbas has of his childhood and of the difficulties his father had
in maintaining the household are always present and underlying his experiences today. Therefore, while the values and practices of the Göreme people
are forever changing, they are always underwritten and influenced by such
memories of a life past. These memories of socio-economic changes in the
village will thus form the second focus of this chapter.

First, though, I will approach the issue of gender by referring back to the
large number of Esin's photographs that depict views of Göreme village. The
point that those photographs raise is the importance of Göreme the place in
villager identity, both male and female.

Göremeli/people of Göreme

Göreme is the place where Esin was born and where she lives. It is therefore not surprising that she depicted her life through so many scenic views of Göreme. Moreover, the photographs she took might also reflect a positive re-evaluation among Göremeli of the 'Göreme landscape' that has been promoted largely through tourism. As I have already noted, there is a sense in which local people can adopt a 'tourist way of seeing' and valuing the place in which they live and work, and Esin's photographs convey her awareness of Göreme's landscape as somehow special. Contrary to being 'peasants without pride', as depicted by Schiffauer (1993) in his discussion of the effects of outward migration on Turkish village life and villager identity, the people of Göreme have retained, or even gained, a sense of pride in living among the ancient and picturesque 'fairy chimneys' that have become world-famous objects of tourist and academic interest.

'Göremeli' refers to a person being of or from Göreme, and is the status accorded a person through birth within the village or through male lineage connected with Göreme. Anyone from outside the village is a *yabanci* (stranger/foreigner), though visitors to Göreme from outside Turkey have, according to villagers, an overriding identity of *turist*. In an interview with Ali, I asked what it meant to be Göremeli:

> Göremeli…It is where I was born, where I grew up. This is my land, this is my last stop if I die. I don't own the place, and so I won't sell the place either. This is my town, I mean, you are a foreigner here, so you can't say 'this is my land', because you don't have ground here. Like my grandfathers, my father's father and my mother's mother, all their lives they lived here and they struggled here, more than me. And it is not for me to leave them behind. Do you understand? I go to their grave, and I go to their garden, and I don't want to leave empty, what they left behind. I don't want to sell their house, I don't want to sell anything, because it is what my family left for me…And this means something for us, like I can't explain it, something inside, you have it inside.

Belonging to Göreme is the most important aspect of villagers' identity.[3] To quote Stirling,

> People belong to their village in a way that they belong to no other social group. On any definition of a community, the village is a community.
>
> (Stirling 1965: 29)

It is important, however, to consider the connection between gender and the issue of identity and place.

Since lineage is entirely patrilineal, the identity of men is more strongly embedded in place of birth than is the identity of women. Though marriage is ideally endogamous, it also takes place between villages, and so a woman is just as likely to live the latter part of her life in another village as she is to remain in her birth village. Moreover, a bride becomes the responsibility of her husband and his family, and where the bride joins a family of another village, she should then consider herself to belong to her new village. In practice, this ideal is not always the case. When I asked women who were married into Göreme if they considered themselves to be Göremeli, some answered that 'yes, they were now Göremeli', while others replied that they were of the village in which they had been born, although if other villagers heard such a reply, they urged the incomer to 'let go' of her home village. In marked contrast to men, these mixed responses show a certain ambivalence among women concerning their sense of membership of a particular village, and in turn their identity in connection with place.

The geographical grounding of identity has implications regarding notions of insiders and outsiders, 'us' and 'others'. That is, the different basis of women's and men's identity is pertinent to the topic of tourism because it influences their various ways of viewing and relating to 'others' – to tourists. For women in Göreme, identity relates more to their being Muslim than it does to their belonging to the village, and so religion is the stronger axis upon which a sense of 'us' and 'them' turns. For men, on the other hand, that axis lies most definitely on the village boundary and on being Göremeli.

The issue of gender is fundamental to an understanding of social relations and organisation in Göreme village, and thus to tourism and tourist interactions there. As previously noted, a pattern of gender segregation and patriarchy underwrites the duality that exists in Göreme life between *türizm* and *kuru üzüm*. The next section will consider why and how that duality is upheld, and how it is lived out by both the women and the men of the village.

Gender segregation: the duality in Göreme life

Gender and power in the household (ev)

Social organisation within the Göreme household is patriarchal, and it is also hierarchical in relation to age. The eldest male within the household (*ev*) carries an all-encompassing and unquestionable authority, an authority over the younger males who in turn hold authority over their wives and children. In many respects, this authority is implicit, since it is generally clear what the daily duties of the female members of the household are, and the older women are able to check the younger in this regard. The authority of the patriarch holds regarding all major decisions to be made concerning his household. These decisions include: the size of his household (that is, if and when his sons should break away from their natal home to begin a new household); the marriage of his daughters and granddaughters; and all economic activities connected to the household.

The concept of 'permission' (*izin*) is central to the authority that men hold over 'their' women, because it is the responsibility of men to bestow or not to bestow *izin* on their womenfolk.[4] One of the major ways that male authority visibly manifests itself through the concept of *izin* is in relation to the issue of movement outside the home, summarised in the Turkish verb *gezmek*. This in turn links with the principles of 'honour' and 'shame' that have been widely associated with each other in anthropological literature concerning Mediterranean and Middle Eastern societies.[5] In an interview, a village girl explained to me the meaning of the concept of *gezmek* for women in Göreme:

> They can only comfortably *gezmek* in their street, they can't *gezmek* in the centre. Maybe if they want to get somewhere, they can pass through the centre, but they cannot sit and drink tea, or sit and eat – they can't do anything. If they go out, it is always to a house – to their neighbours' house – to their friend's – always like that. They can *gezmek* comfortably like that only, nothing else.
>
> [*trans.*]

I asked why this was the case:

> They think that maybe if she goes to the centre another boy will look at her, they are jealous, do you understand? But nothing will happen, I mean, they will only look, it is not a problem, but that is the way Turks think. Not everywhere – in Göreme they think that way. In Göreme women cannot *gezmek* comfortably, it is not possible, otherwise their husband will be angry, he will always be angry. They'll ask, 'Why does she *gezmek*?', 'What is she going out for?' It is necessary for them to ask their husband for everything they do, everything they do.
>
> [*trans.*]

A girl or woman must always be granted permission by her male counterpart if she wishes to go anywhere outside the immediate neighbourhood – to the market, a wedding, to visit relatives, and so on. As I indicated in the introduction chapter, these village codes for women's conduct also impinged on my behaviour and my ability to conduct research in the village. I was regularly told by villager friends that I *gezmek* too much and that I should stay in the house more. Abbas, my 'uncle', frequently shouted at me '*Köpek gibi geziyorsun!*' ('You wander around like a dog!'). I increasingly felt the necessity to ask for *izin* when I wished to leave the village, to go to town or to walk or cycle out to the valleys, and this provided me not only with a sense that I was treating my adoptive 'uncles' and 'brothers' with their due respect, but also with a sense of protection. For if I had correctly obtained *izin*, whoever had granted the permission would then hold a certain responsibility for my safety while I was away. The longer I stayed in the village, the more I had to accept the 'protection' afforded me, along with the male 'authority' over me that this implied. In Chapter 7, I discuss these issues further in relation to tourist women.

For an understanding of the symbols and beliefs underpinning gender relations in Göreme society, it is useful to draw on Delaney's ethnography *The Seed and the Soil: Gender and Cosmology in Turkish Village Society* (1991), and also on Marcus's discussion of 'Islam and gender hierarchy in Turkey' (1992). Delaney highlights the roles of man as planter of the 'seed' (*tohum*), and of woman as the 'field' (*tarla*) in which the seed is planted, in Turkish beliefs about procreation. As the maker and planter of the seed, the man is the generator of life and this is what grants men their power and authority. However, his honour depends on his ability to guarantee that a child is from his own seed, and that in turn depends on his ability to control *his* women.[6] Simultaneously, a woman's value arises from her ability to guarantee the seed of a particular man:

> The value of a woman depends on her virginity before marriage and her fidelity after marriage; this is socially recognised by her conformity to the code of behaviour and dress.
>
> (Delaney 1991: 40)

This is complicated, however, by the further belief that 'unlike male sexuality, women's sexuality is located within a body beyond the control of the mind' (Marcus 1992: 83), and so women's modesty can only be accomplished by externally imposed restraints:

> Women are unable totally to control their bodies through the exercise of the power of the mind,…men must therefore control women if the community is to retain the moral order based upon the clear separation of two genders. Community order therefore rests upon male control of women. This point is crucial in understanding gender relationships in Turkey.
>
> (ibid.)

In Göreme it is shameful for a woman to *gezmek* because to wander around in public spaces is to put herself in a position where she may be 'looked at' by men other than those belonging to her household. If she remains always within the house (or the immediate neighbourhood), she is protected from shame (*namus*). Simultaneously, a man's reputation or honour (also referred to as *namus*) is protected if the shame of his womenfolk is prevented, and so in order to protect his honour he must assert his given authority to prevent the shame of 'his' women.

The above is the normative discourse, representing the ideal that maintains the general *status quo* in Göreme society. In practice, girls and women do contravene the norm to varying degrees, and in so doing soften and undermine male authority over them. Girls might wear jeans and T-shirts throughout the day and then quickly change into the traditional *şalva* (baggy trousers and headscarf) before the male members of the household come home in the evening. Daughters may assert their wishes regarding who they do or do not want to marry, and if they argue long enough they may get their own way. Occasionally, boys and girls meet illicitly, this being possibly the riskiest of all secret activity. Other examples include the 'naughty' (*yaramaz*) behaviour girls and also their mothers might engage in while supposedly working, hiding behind vine bushes out of sight of the authoritative gaze of the accompanying male and lazing around eating grapes and cracking open walnuts instead of picking the crop.

Since women spend most of their daily life with other women and apart from men, their relationships with each other are extremely strong, and their psychological dependency on men limited.[7] This very close-knit nature of the 'community' of women is tolerated, in its formal state at least, because of the group work undertaken by women in relation to agricultural and food-production activity. However, the separate women's community is ultimately threatening to men, and can cause men to feel ill at ease; if given the opportunity, men are quick to disperse a group of chattering women. Men's honour is indeed highly vulnerable, and it is because this honour is largely dependent on women's behaviour that men must control women and ensure their separation from what might be regarded as the 'public' domain of society.

Male authority is thus inextricably linked to gender segregation, and the concepts of honour and shame (*namus*), *gezmek* and *izin* are the lived manifestations of an ideology of male control. These issues are highly pertinent here in their links with the way that tourism has developed and taken shape in Göreme village, specifically regarding the different ways that men and women are involved with and related to the tourism processes. This connection becomes clearer if we now continue the discussion by looking at the particular ways that gender segregation is manifested spatially.

Gender and separation

Göreme village is divided into residential quarters (*mahalle*) that surround the village centre (*çarşı*). Each *mahalle* forms a roughly bounded and loose-knit community within the village community. Men often explained their friendships and business partnerships by saying 'I know him well because we grew up in the same *mahalle*'. For women, the *mahalle* is where they spend almost all their time, and their female neighbours are therefore those with whom they form their closest relationships. Only if they have relatives living in another *mahalle* would women visit another quarter (here I am referring to women and girls beyond puberty, as previous to puberty girls are free from the restrictions of modesty). A woman would never, except perhaps occasionally for a wedding, go to quarters of the village where she does not have any relatives. Within their own neighbourhood, on the other hand, women may move around quite freely.

They often visit each other's houses, and groups of women often sit in the afternoons either on their doorsteps or outside their houses in the wider, more open parts of the mostly narrow streets.

Women are by no means secluded in Göreme society. They are not hidden away inside their homes, they are not banned from the central Göreme streets and they are not harassed if they enter male space, such as the market place. However, men and women do have their clearly separate domains. The women's domain – social, economic and religious – is centred around the household and the immediate neighbourhood, while the male domain is the *çarşı*, symbolised by the main mosque, the market place and the tea house.

The tea house is most definitely male space, and it is the place where men pass their spare time playing cards or backgammon and discussing village affairs and politics. It is notable, however, that the tea house in Göreme has been increasingly transgressed by tourist women, who sit looking rather uncomfortable among the village men. Obviously keen to attract more tourists, in the summer of 1998 the tea-house manager employed a Dutch woman as a waitress. The sight of this young blonde woman serving tea to elderly men in old suits and flat caps was startlingly anomalous.

Women may pass through male space: they may go to the weekly market and they may attend village ceremonies, such as children's concerts and national day celebrations, but they have no 'place' there. Likewise, men have no place in the household during the day when they should be out at work or in the tea house. Abbas, for example, spends most of his days and evenings during the summer either working or going around visiting his friends (*gezmek*). In the evenings, men do also sometimes partake of formal 'visiting' with their families, though they do so more often in winter because they are busy in their tourism businesses until late in the evenings during the summer.

Women's relationships to different sections of the village are symbolised by the amount of head-cover they wear in different places. When inside the home during the day, when no men are around, women wear the small headscarf (*yemeni*) typical of the region. Outside the home, but within the immediate neighbourhood, they still only wear this small headscarf, although more securely than in the house. When they leave their immediate neighbourhood, they secure the small scarf so that it covers their chin and mouth, and they put on a second larger scarf (*çarşaf*) that drapes down to cover the full top half of their bodies. When they go to work in their gardens, which mostly lie outside the residential village area, women might remove their outer scarves again, as if re-entering the home sphere. In general, women are quite free with their behaviour and dress while in the fields, indicating that the fields are, through their absolute association with the household economy, considered part of the domestic sphere. This may gradually change, however, as the walking paths of increasing numbers of tourists exploring in the valleys pass close by many such fields.

All women wear the outer scarf if they pass through the centre of the village or visit the market place, and unmarried girls should follow the same pattern from the age of puberty, though this depends largely on how strict the male members of their household are. When permitted, many younger women today wear a more 'modern' form of headscarf that is considered fashionable in urban areas of Turkey. A few girls and women may be seen without any headscarf, and they are the daughters of returnee migrants who, having been educated and urbanised, are considered to have now moved beyond the usual social codes of Göreme life. They may be criticised for being 'open' (*açık* – their hair is uncovered), but are generally not subject to as much critical gossip as a village girl would be.

Similarly, tourist women are considered 'open' in contrast to the 'closed'/'covered' (*kapalı*) women of the village. Tourist women often incite much discomfort and criticism in this regard, though the threshold of criticism is gradually being stretched so that today it is their wearing short trousers and mini-skirts, which is considered offensive, rather than their uncovered hair.

So the degree of cover a village woman requires depends on which general domain she is in. She should only pass through the centre of the village when it is necessary to do so in order to reach the fields, or to go to the market or the health clinic. She should always pass through quickly and, if possible, should take one of the smaller streets to avoid walking along the main central road. One day during the first summer of my fieldwork, the *Belediye* (municipality office) announced the move of the site where the weekly market took place because of congestion in the old site next to the bus station. That site was conveniently close to the main residential areas of the village, but the market was to move to a more spacious area available at the far end of the main road. As the village loudspeakers repeated this announcement from the *Belediye*, the women complained and cursed the mayor for his decision, which would now make it difficult for them to walk to the Wednesday market. Not only would the new site be too far to carry goods back from, but it would also now involve passing to the other side of the village centre and their men would not allow that. Husbands and sons could buy the food items they needed, but how could these men choose the material and thread needed for the women's embroidery? Indeed, the villagers quickly adapted to the new market place, although the women from that *mahalle* no longer visit the market every week, and they always take a small side road rather than walking along the main road cutting through the village centre. As one young woman who married into the village told me:

> That road [the main road] is dangerous for me. I don't use that road much. There is a lot of traffic and I feel depressed there. There are restaurants with people eating there, and I feel uncomfortable. I feel as if they all look at me. They don't look at me, but I feel that way and I feel shy.
>
> [*trans.*]

The centre of the village is not only the male domain, but also these days the main tourism area of the village. Consequently, when women do pass along the main road (for example, riding on a donkey or cart on their way to the fields), they often find themselves to be the unwitting photo opportunity of tourists. The many tourists sitting on restaurant terraces or hanging around in street cafés love to take a chance snapshot of these 'quaint' scenes of veiled women riding on donkeys, but this only adds to the general discomfort felt by village women in this realm, and so encourages the further separation of the two gendered realms in Göreme.

The social and spatial gender separation discussed here thus has important implications regarding women's relations to Göreme's tourism and its developing economy. Women's access to the newly developed economic sphere is not limited directly by an ideology concerning the division of labour *per se*, but rather by an ideology that ensures the separation of the genders. Most of the tourism businesses in Göreme are situated in the triangular area of the village along the main roads, the market or business centre. It is inappropriate for women to spend any time in that area, and so they are largely excluded from the tourism processes in the village. Likewise, in cases where tourism has entered the domestic sphere, where village cave-homes have been converted into *pansiyons* for tourists, a separate house is eventually built for the women and children of the family. This is not because the women should not partake in such work – running a *pansiyon* is largely 'domestic'-type work anyway – but because it is inappropriate for women to be in close proximity to unrelated men.

Before tourism, agriculture was the main source of production and economy, and the household was the main economic unit. Then women's labour was associated with economic production and was highly valued. Now that the economy in Göreme is largely based on tourism, the economic centre has shifted almost entirely to the tourism realm in the centre of the village. Very little of the agricultural produce is currently sold on the market, and so agricultural production has become almost entirely a 'domestic' activity. This is not a process occurring just through tourism, as Delaney notes a very similar process occurring in Turkey generally through the process of 'development':

> In the process of mechanisation or development, men's focus is being drawn outward; women are being left behind, more than ever enclosed in the house...It is at this juncture...that women tend to become identified with the 'domestic' private realm of reproduction and men with the public realm of production.
>
> (Delaney 1991: 267–8)

In Göreme, most of the agricultural work is carried out by women. This work has now become something of a secondary activity that runs alongside the main money-earner, which is tourism.

Gardening and food production

The attachment to agricultural production is shown through the fact that most Göreme households have kept hold of their land even when a large part of their economy is now based on income from tourism. Though food-production activity has become almost the sole responsibility of women, some older men, such as Mehmet in the vignettes above, have chosen to stay with what they know best – working the land. Some villagers get the best of both worlds by renting out their old cave-house, converted into a *pansiyon*, to younger men who have no property of their own, and then farm their land in the comfort that they have an ongoing additional income.

From spring to autumn there is a constant flow of work to be done: turning and weeding the soil; sewing and planting; pruning grape vines and fruit trees; harvesting and threshing wheat; and, last in the season, picking and drying grapes. Much of this work is carried out by hand or using traditional equipment, with the use of horses for ploughing and donkeys for transportation. A few men in the village own tractors that can be hired for heavier work, and during the wheat harvest cutting and threshing machines are hired from neighbouring villages. During the summer months elaborate processes of drying and preserving of foods are carried out in preparation for winter. These processes are aided by the climatic conditions within the cave-houses, which are steady in temperature and humidity level and thus well-suited to food storage. Bulgar wheat, spaghetti and bread will keep for one or two years in the cave cellars. Tomatoes and cabbages are pickled, and apples, pears and grapes will keep moderately fresh throughout the winter.

As autumn approaches, the workload increases. The grapes are harvested and sun-dried, either for winter storage or for selling to local wineries or *rakı* (the national Turkish alcoholic drink) factories. Another batch of grapes is juiced and boiled until it becomes a sticky sweet syrup (*pekmez*) that, before the arrival of beet sugar in the shops, was the main source of sugar. A third activity in the autumn is the making of a huge batch of flat bread that will last each household at least through the winter months. Due to the intense work involved in such tasks, women share the load by working together as a group or community (*topluluk*) of relatives and neighbours. Day by day the women get together and produce all the bread and all the *pekmez* that each household in turn will need. As they work, they joke and gossip, and are able in this way to keep up the hard work from early morning to well into the evening if necessary to finish the task. This co-operation between women is a fundamental part of village life and economy, even though, unlike men's more formalised business partnerships (*ortaklık*), it is not referred to in any recognised way. The women seem to simply get on with it: as bread-making time approaches, women meet with the relatives and neighbours they usually join together with to arrange which days they will work at each other's houses to roll out and cook the bread. They then do this, perhaps over a two-week period, until all the group's bread is done.

Plate 4.2 Making bread to last through the winter is a full day's work, shared among a
group of women.

While the production of the main staples – such as bread, *pekmez* and bulgar
wheat – is clearly an integral part of women's lives in Göreme, along with the
reciprocal co-operation that goes with it, the income from tourism has made it
increasingly possible to purchase most food items from the market. An argument I witnessed between Abbas and his wife Senem concerned precisely this
issue. The previous spring Senem had sowed wheat even though Abbas had told
her not to bother because it was easier now just to buy wheat and flour from
the market. The argument was sparked by the difficulty they were having in
hiring the threshing machine, which was in great demand at that point in the
harvest season. Abbas wanted to give it up: 'Burn the field,' he said. 'It is
cheaper anyway just to buy wheat these days.' For Abbas, who now spends
more time looking after his tour agency, the effort involved in growing, drying
and grinding the wheat was not worthwhile. Senem, however, insisted that they
carry on: 'It is cheaper to produce it ourselves, and in any case it is *adet*
[custom], we are used to doing it.' Abbas's elderly mother joined in the argument, adding that: 'Before, doing the wheat was the only way we could eat.'
The deep-seated feelings of necessity towards garden work and food production
continue, even though they are no longer an economic necessity. Despite the
significant social and economic change that tourism has brought to the village,
an element of continuity exists, particularly on the part of the women. The

memory still lives on of times when survival of the household depended directly on working hard in the gardens, and it seems certain that such memories of hard winters and empty stomachs will die hard, even though those times now seem to be firmly in Göreme's past.

Memories of a life past: social change in Göreme

Hard times

One evening I sat together with three generations of women and girls and listened to an elderly neighbour's stories about her earlier life. Everyone listened eagerly, often laughing at the sometimes ridiculous and pitiful antics being described of the times she and her husband used to have to stay out for days on end in fields far away from the village. I asked the younger women if they often sat and listened to the old people's stories in that way.

> Yes, the old people often talk like that about when they became a bride (*gelin*), when they became a mother or a father. They tell of how hard it used to be for them, how they all slept in one room, and how it was dark and cold in the winter.

Social and economic change in Göreme has been profoundly influenced by the development of tourism business in the village, but, as was noted in the introduction chapter, they are also the outcome of national economic reforms effecting change in economy and culture throughout Turkey. These reforms included the external migration programmes that occurred at their highest level in the late 1960s.[8] Indeed, Stirling notes the incredible scale on which demographic, economic and social changes have taken place since Turkey became a nation-state in 1923. His own experiences and longitudinal studies in two Central Anatolian villages were witness to a sharp increase in the standard of living:

> GNP per capita increased roughly threefold, 1950–1986. What I see among the villagers and their urban descendants makes these calculations a plausible index. Food, clothes, heating, housing, transport, health services, household durables and furniture, consumption for pleasure and for display, operating capital for farming, investments in agriculture, real estate and businesses are all incomparably more plentiful per person than in 1950.
> (Stirling 1993: 7)

In Göreme, everyday comments and stories told alerted me to the very deep-seated temporal nature of people's experiences of their lives in the village. While tourists – and anthropologists, for that matter – tend to come and go, the people of Göreme have a deep attachment to the place and to each other, an attachment perpetuated through the telling of anecdotes of happenings together, and through referring to memories of times gone by. Experience in the present is layered on memories of the past, and these memories include those of the change that has occurred at an increasing rate, particularly over the past few decades. This change is shown through the memories of people like Abbas and Mehmet, and even older members of the community such as 'Anne', and these memories of how life used to be are constantly referred to in daily life, serving as reminders of the need to continue to struggle and take risks in order to secure a better future. Abbas often remarked to me how his father had been unable to maintain a living within the village and so had gone with his wife to live on the mountains to tend sheep, leaving Abbas and his brothers to stay with their uncle in the village while attending school:

> There were no jobs to do. We had only a small garden. We were a big family, with five children, but with few gardens. He bought sheep and stayed with them on the mountain. Then he went to Nevşehir to sell yoghurt, and he sold fat. For his five children it was necessary that he did that.

The younger generations are regularly reminded by their elders that if they do not work hard enough, they will not have money to buy tea in winter. Any sense of nostalgia seems always to be limited by an overriding memory of the hard times when the struggle through winter was real and the ability to purchase items such as tea or shoes was always precarious. This memory, and the sense of relief that tourism has brought, was shown to me one day in spring when I went with Abbas and his wife to their vine gardens to prune the plants ready for the summer growth. They told me that the plants had been damaged by the late cold weather in April and so would not produce many grapes either this year or next. Abbas then remarked, 'If there was no tourism, this would be a big problem. Maybe we wouldn't be able to buy any bread in the winter.' These days, though, since tourism has brought relief from the absolute necessities of agricultural production, this was more of an annoyance than a big problem.

While Abbas and his contemporaries were growing up, a communication infrastructure was being developed that promoted a general climate in which men would begin to look elsewhere for work. In her comparison of studies of Turkish villages from the 1940s and 1950s, Keyder also notes:

> [In the 1940s] The village was often the only life-world, and all national and world concerns were filtered through its structure; market transactions were few and infrequent; most of the output was for the household's own consumption or for local exchange...By contrast the literature of the 1950s is full of descriptions of market adaptation, new inputs and changing technology; the importance of banks and other formal credit institutions, and, of course, the beginnings of urbanisation. In the 1950s the Anatolian peasant seems to have resolutely started on a road towards commercialisation of outputs, of inputs, and of his own labour.

(Keyder 1993: 171)

During the 1950s and 1960s Göreme men began to leave the village to seek cash-earning opportunities. Some were less successful than others, as is shown above in Mehmet's unlucky attempts both to migrate to Europe and to gain employment in a textile factory in nearby Nevşehir. Abbas had also tried unsuccessfully to find employment in Ankara when he was in his twenties. Others did manage to find work in Nevşehir as mechanics, or they became civil servants for the gradually stabilising government in nearby towns, as Mehmet did. Many others succeeded in migrating either to urban areas of Turkey, such as Ankara or coastal cities such as Mersin, or to the northern European countries of Germany, Belgium and Holland. By 1970 as many as a hundred Göreme families had migrated to Europe and many others were working in haulage and transportation having used money earned in Europe or credit from the government to buy their own trucks. By 1980 more than seventy large trucks were owned by Göreme men, some of them being used to transport goods between northern Europe and the Middle East (Bezmen 1996).

Migration and the making of Göremeli 'outsiders'

Agreements between the Turkish government and those of certain northern European nations continued throughout the 1960s and early 1970s until the migration programmes were brought to an abrupt halt with the international energy crisis in 1973. The migration agreements evolved throughout the intense migration period (Abadan-Unat 1986). Initial ideas concerning Turkish migration were largely based on men's temporary employment abroad. However, as the need for workers remained prevalent, new policies were developed whereby many migrant workers were given citizenship and allowed to be accompanied by their wives and families (ibid.). In such cases, of which there are many from Göreme, women also became migrant workers, employed in factories – often alongside their husbands.

There are also many cases, however, where the wives and children stayed at home in Göreme. Some remained in the house of their parents-in-law, but besides those the situation arose whereby women became the acting heads of their households and took up the new position of being not only in sole charge of family agricultural production, but also of financial matters and the task of raising the children. This situation may have worked towards increasing the status position of village women, though I was made aware of a few examples of women now considered 'abandoned' by husbands who have not returned.

Many of the Göreme people who initially migrated to Europe have now returned, while others return for long periods during the summer and to be with family during religious holidays and the months when marriage takes place in the village. Whether permanently returned or not, many such families have used their savings to build 'modern' homes in the flat areas of Göreme village away from the caves. Some permanent returnees have also invested in tourism business in the village.

The consequences of internal and external migration on villagers' lives and the socio-economic situation of the Göreme community are of course many and varied. One consequence was the creation of obvious inequalities in wealth and corresponding status among villagers, which has led to bitter displays of gossip and criticism, particularly from those who never managed to leave the village to work. The people who left the village are accused by those who stayed of opting for a much lesser quality of life working in a European factory, and suffering the renowned racism of northern Europeans, just so that they can show off their accumulated wealth in an 'empty' and 'uncultured' manner. Indeed, the wealth of returnees is often displayed in the form of Mercedes cars or northern European-style furnishing in their modern homes. As one Göreme woman, who keeps cows in her cave-home and whose husband runs a cave-home *pansiyon*, told me:

> The people who went to Europe come back with all their money, build new houses and hold their noses in the air thinking that they are better than us just because they've got rich. But they're not. They think that more money means more *insanlık* (humanity, goodness), but it is not connected. You can't buy *insanlık* with money, can you?
>
> [*trans.*]

Concurrently, returnee migrants might accuse those who stayed behind of remaining uneducated and uncultured villagers/peasants. Indeed, young men and women whose parents had migrated away from the village are far more likely to have attended university. One returnee migrant told me:

> It was not easy for me to live here because I grew up in Holland. The culture was different for me – OK, I know I speak Turkish but not that well – because I grew up in Holland.

He went on, however, to tell me how good the life of his parents is now that they have their retirement pensions from Holland.

Other stories, in contrast, conveyed the great difficulty many people have in returning to the village. One man, who had lived and worked in Germany for ten years and now owns a tour agency in the village, told me of how he had not been accepted into the village tea house for a long time after his return. A man's sense of belonging to and place in the village clearly depends on his regular attendance and ability to partake in village discussion at the tea house, and is often referred to by returnee migrants in illustration of the difficulties they face in regaining their social position within the men's community. The worst scenario for a Göremeli man is to walk unrecognised into the tea house because of his being away for so long. Interestingly, men's working in tourism can have the same repercussions; because of their spending most or even all of the day in their tourism business, they effectively 'leave' the village during the summer months and, come winter, find it difficult to re-enter the tea house. On this, a young man who works in a carpet shop told me:

The people who moved to Holland and everywhere, their sons are outsiders now. And maybe we'll end up like that too, because we don't fit into our village anymore.

Migration has led to a situation where people are both Göremeli and, at the same time, 'outsiders'. It has blurred the boundaries demarcating the Göremeli community, and it has concurrently forced the villagers to reconsider their place in the world and to open their lives to 'other' possibilities. The people of Göreme have inevitably gained links with a much wider world than they used to have before migration occurred. As Abadan-Unat states in reference to Turkish rural society and outward migration,

> The concept of distance has changed…in the 1930s men walked for two weeks to find work in Ankara. Today Turkish workers are employed on five continents. Peasant women in *şalva* and men with traditional baggy trousers are boarding planes for Stockholm, Sydney, Berlin. For them the sphere in which one earns one's bread has grown to a global scale.
>
> (Abadan-Unat 1993: 207)

There is today, as there has been for some decades in Göreme, a significant amount of movement in and out of people, communications and goods. With this movement has come a striking awareness, even if only in people's imaginations, of ways of life and opportunities that exist outside Göreme, and indeed outside Turkey. Although they may have never been as far as Ankara themselves, villagers invariably have relatives in northern Europe and have heard from them of the economic possibilities there. Many still hold out the possibility of their sons' employment abroad, even though in reality their sons' employment in Göreme's tourism might prove to be more gainful. As Schiffauer (1993) has argued, outward migration has caused Turkish villagers to lose pride in their village life because of their perception that a future outside the village is better than to remain; those who do remain in their village are considered 'feeble' and 'losers'.

While, to some extent, idealising a future outside the village has also occurred in Göreme, it has been limited by the crucial difference between Göreme and most other villages in central Turkey. That difference is, of course, that the men of Göreme have been presented with an extra, viable economic opportunity within the boundaries of their own village: tourism. In the next chapter, I will go on to discuss the advent and development of tourism in Göreme, and the changes it has brought about in villagers' lives. That chapter will thus build on this one, in which the social, spatial and seasonal duality in Göreme life has been described, forming an understanding of the parallel but quite separate lives of men and women, as well as the simultaneous processes of continuity and change.

Notes

1   For ethnographic accounts of patriarchy and gender segregation in Turkish rural society, see Bellér-Hann and Hann (2001), Delaney (1991, 1993), Sirman (1990) and Stirling (1965). For more general discussion of gender in Turkish society, see also Abadan-Unat (1986), Kandiyoti (1991, 1996) and Marcus (1992); for comparative discussion of other Muslim societies in the Near and Middle East, see Abou-Zeid (1965) and Tapper (1991).

2   Indeed, it has been made clear that is important to consider gender issues in discussions on tourism generally (Kinnaird, Kothari and Hall 1994; Sinclair 1997; Swain 1995).

3   As Stirling has pointed out in reference to Turkish rural society, 'the village itself is the most striking social group' (Stirling 1965: 26), so the particular village to which individuals belong is the primary factor in their identity.

4   In her discussion of systems of authority in Turkish villages, Delaney (1993) also points this out.

5   See, for example, Blok (1981), Boissevain (1979), Bourdieu (1966), Campbell (1964), Gilmore (1987), Herzfeld (1980), Peristiany (1966), Pitt-Rivers (1966) and Tapper (1991).

6   It should be noted here that Peristiany's initial contention that an 'honour-shame' complex exists in a similar form and with similar reason throughout the Mediterranean region (Peristiany 1966) is much debated. It is doubtful, for example, whether the concepts of honour and shame can apply in the same general form across Islamic and Catholic societies. Similarly, it has been argued that Islam itself varies contextually and cannot be focused on as being the sole cause of gender ideologies (for example, Abu-Zahra, cited in Goddard 1994).

7   Marcus (1992) has also made this point in reference to gender relations in Turkish society.

8   Turkish outward migration is the focus of many studies: for example, Engelbrektsson (1978), Kocturk (1992), Stokes (1992) and the articles by Schiffauer and Abadan-Unat in Stirling (1993).

5   A community in competition
The business of tourism in Göreme

'Göremeli' means – someone who keeps the old traditions and customs of the village, and someone who respects the elders and loves the youngsters. And who helps each other and loves each other. But unfortunately, from tourism, it's changing a little bit. Göremeli is going away from this principle. When there was no tourism in Göreme, people had a village life, so the main income of the villagers was from farming, and so...they had time to talk with each other, so the relationships were much stronger. But when tourism became so important, people do many jobs, like *pansiyons* or whatever, so for money they would easily disturb each other even though they are family, you know, even though they are members of the same family. So this changed the people's lives.

(Göremeli carpet shop owner [his own English])

Significant changes have undoubtedly occurred in the lives of the Göreme people with the move from farming to tourism. We saw towards the end of the last chapter how Göremeli people, like those of other Turkish villages, began during the 1960s and 1970s to act to improve their lot; to remove themselves from the hardships associated with living mainly from agriculture. Many villagers migrated at that time, imagining a more prosperous future for themselves outside the village. Tourism has to a large extent worked to reverse that process, returning a sense of pride and future within the village. Young men who were otherwise likely to have gone to work elsewhere, remain and are working in the village; others who had already migrated out are drawn back. The chance and hope of having a prosperous future on home ground has been greatly increased because of tourism development.[1]

Simultaneously, however, that chance of prospering from tourism is constantly threatened by both external and internal influences. Threats from outside include large tourism companies, as well as would-be entrepreneurs from other parts of Turkey or abroad, who would like to get in on Göreme's tourism opportunities. On the inside, as the carpet seller quoted above states, the threat is that the Göremeli men have been thrown into overt and intense competition with each other as they battle among themselves for the custom of tourists.

On the one hand, then, there is the issue of 'community-participation' in tourism development, and the relative strengths and relationships between insider and outsider players. It was noted in Chapter 1 that the nearby town of Ürgüp provides a useful comparison to Göreme in this regard. While Göreme has been protected because of its association with the Göreme Open-Air Museum and National Park, Ürgüp has seen the development of a large-scale tourism sector as national and international hotel chains and tour operators moved in during the 1980s, taking advantage of the incentives provided under the Tourism Encouragement Act. Many of the smaller businesses owned by Ürgüp people were consequently forced to close because of imperfect market competition, creating a situation of 'unsustainable tourism at the local level' (Tosun 1998). It has frequently been argued that small business development, together with a certain level of co-operation among stakeholders, makes for successful or 'sustainable' community-based development.[2]

The necessity for some sort of co-operation leads to the other issue here, and that is the question of what happens to the notion and ties of community when people become increasingly involved in capitalist economies and markets. Bellér-Hann and Hann (2001) discuss this issue in relation to the people of the Lazistan region in north-eastern Turkey, noting that because capitalist economies are predominantly individualistic, the growth of free enterprise in Lazistan as a result of Turkey's market liberalisation in the mid-1980s stimulated competition and selfish acquisition.[3] Bellér-Hann and Hann also note in their ethnographic study of the development of a market-based economy in that region, however, that state top-down pressures together with market liberalisation are always mediated by some sort of principle of community.

This chapter outlines the processes of tourism business development in Göreme, emphasising how, because that business has been developed by the local villagers, the pattern of growth has been effected by certain aspects of Göreme social relations and behaviours. The chapter then goes on to consider the extent of the ensuing changes to villagers' social relationships and behaviours due to increased competition and individualist enterprise.

Getting into tourism

Before tourism came many people went to Germany to work...And Göreme started to become known and famous in the world from the visitors to Göreme Museum, and there were just few tourists coming to Göreme. There were, in 1965, just one or two *pansiyons* in Göreme. And their neighbours saw that they were making some money. And when you have no job in town and you probably don't get much from your field, you think that you will also make a *pansiyon* in your home or hire a larger home. And so the number increased every year, and so they learnt tourism business from their friends, from each other. And the ones who went to Europe or to bigger cities like Ankara saw that life was changing in the village, and so they decided to turn back to their village.

(Göremeli carpet seller [interview carried out in English])

During the 1970s and early 1980s the number of tourists passing through the village, then named Avcilar, in order to visit the nearby Göreme Open-Air Museum was becoming substantial enough to make tourism a viable economic option for would-be local entrepreneurs. It was seen in the vignettes in the last chapter how Abbas and his brothers were some of the first in Göreme to get into the business of tourism in the village. In the late 1960s and early 1970s, as teenage boys with an entrepreneurial spark, they had the initial ideas of how to make a bit of money from the first few tourists who were beginning to pass through. Notably, these individuals set the ball rolling and, if we follow Ardener's notion of the 'event-rich' character of remote areas, these few early entrepreneurs had a great deal of influence in Göreme's tourism:

> Event richness is like a small-scale, simmering, continuously generated set of singularities, which are not just the artefact of observer bias...but due to...the enhanced defining power of individuals.
>
> (Ardener 1989: 222)

In small societies such as Göreme, in other words, the actions of individuals are more noticeable and consequently influential than in larger urban areas. As we shall go on to see, the growth and shape of tourism business in Göreme has followed this principle, with new practices and ideas being quick to spread.

To consider the influence of individual men in the early days of Göreme's tourism, the following are four men's narratives concerning how they began to get into tourism:

Abbas (aged late forties)

When I was fifteen, I sold cards to the tourists. That was at the beginning of tourism you understand...That earned good money. And then I said to myself, if I sell things to tourists, I can earn money. So later, I bought cards and books and started to sell them – and then I sold antiques together with my friend. We did that job, from '68 to '74. In 1975, I went to the military service. After I returned from the military, I opened a restaurant, Göreme's first restaurant. I opened it in 1977. I ran restaurants for about eighteen or nineteen years, and then I opened a tour agency. [*trans.*]

Omer (aged late forties)

When we were school students, we were selling postcards in our pockets to tourists up near the churches over there. There's a restaurant over there, we were living in a cave in our holiday time, this was the start,

---

started selling postcards, it was really good. After I finished school I worked in a factory, a textile factory in Kavabak. And after two years I thought 'it's not the work for me' – it was hard really, and I didn't like it much. So, then I came here and I worked in the municipality office for two years. And then I thought I have to do my own business, so I started the travel agency – and that's thirteen years now I've had this business. And now I also have a hotel and a cave *pansiyon* which I bought last year and restored, and I have some partnerships in other businesses as well.

I asked Omer how he had the idea of opening a tour agency in Göreme:

Well, there were some other travel agencies in Ürgüp, so I checked them and they were doing good, so I said this is the kind of business for me and so I opened my own.

I then asked if there had been enough tourists to run a tour agency in Göreme thirteen years ago:

Yes, there were enough. I was the first one to have a travel agency in this town, so there were enough, because I was the only one, and then the next came, and another came and another.

Tuncay (aged late forties)

I was born in Göreme in one of the caves, and I studied primary school in Göreme and I went to Nevşehir for high school. After I finished high school, I was a local guide for the churches, showing the fairy chimneys to the visitors. And later, when I was twenty-two, I went to Istanbul to do my military service. And after I finished my military service, besides my guiding, I was also interested in carpets. This was our family work, my mother and sisters used to make carpets, and so I decided to open a shop, and I am still running this business, for more than fifteen years now. And I'm enjoying staying in this business, and in the future I'm thinking of expanding my business to outside of Turkey, maybe in Europe or America. This is my dream.

These three men, all in their late forties, are representative of those who first started tourism business in the village. The next narrative is typical of the younger men who have not yet gathered the capital to own their own businesses. In contrast to the older men depicted above, this younger man works as an employee for others and therefore sees tourism more as a 'job':

Hüseyin (aged early thirties)

I try to do different things every year...and then I can get to know people, I can get to know the jobs, what kind of jobs are important, like what belongs to tourism. Tourism is not easy. People think that it is easy, but every job has its own difficulty, in different ways...

I went to school for five years. I finished primary school and I didn't go to high school, I just started working. Before I was eleven, I did three years at the *sanaye* (industry centre) decorating trucks, because tourism hadn't really started at that time...My father was a farmer...We didn't get rich, we didn't get poor either. We just stayed the same.

I started in tourism when I was eleven. I washed dishes for three months at Kaya Camping underground restaurant, and after I started at Rock Valley Pansiyon as a chef. You do everything there, when you are in a *pansiyon*, you must do everything; you must speak the language, you must cook, you must make the beds and wash the showers, and greet the people and check the people to see if you can get them to go to the carpet shop, and you can try to sell the tours before they go out, and that kind of business. When you are in the *pansiyon* position, you must know all these things.

Most of the earlier tourist activity conveyed in these narratives was centred around the museum site of the Göreme valley, situated a mile or so outside of the village. Inside the village itself the idea of tourism business was slow to catch on, with most of the early contact with tourists taking place at a little restaurant and camping stop called *Hacının Yeri* ('Haci's Place'), set in a 'fairy chimney' on the road leading into the village, which was run by an Ürgüp man called Haci. Some of the young men of Göreme, including the three older of the men above, worked at or hung around the museum area and Haci's Place, and it was not until later, when a tourist presence in the village was becoming more strongly felt in the early 1980s, that they decided to follow up on their teenage antics of guiding and selling postcards.

All of those men, now running their own businesses in Göreme, remember the earlier days when the tourists were 'crazy hippies' driving through the area in camper vans:

> Thirty years ago, the tourists who came here were all hippies – they all took drugs and older French women came for sex, and homosexuals too.

Tourists were quickly associated with drugs and sexual immorality, and so working among them was considered far from being a respectable occupation, especially according to the older members of the community. Abbas remembers the old people saying:

> Don't go near the *giaours* (infidels). If you do, they'll cut you up into little
> pieces and throw you off the hill into the valley.

Some elements of these early attitudes towards tourists remain to this day,
although most villagers, particularly those working in tourism business, have
become a great deal more used to tourists. While many villagers do still use the
term *giaour* for tourists, I often observed people being reprimanded for doing so.

Coupled with the suspicion and fear of the 'outsiders' and 'infidels', there
was and still is a tendency to ridicule any innovation. In relation to his opening
of the first 'tourist' restaurant in Göreme, Abbas told me: 'People said to me
"What on earth are you doing that for? Who will come and eat there?".'
Indeed, Stirling noted what he regarded as a 'proverbially conservative' character of Turkish villagers, stating:

> If people take, as they are almost certain to do, the view that the innovation
> is malicious, pretentious, dangerous, impious or absurd, the innovator, if he
> persists, has to face criticism, ridicule or even ostracism.
>
> (Stirling 1965: 291)

This tendency might be attributed to the characteristic in peasant societies
described by Bailey as a constant 'competing to remain equal':

> Skills and energies go into keeping people in the place that they have always
> been: they run hard in order to stand still. It is the kind of world that
> stamps heavily upon change and innovation.
>
> (Bailey 1971: 23)[4]

In Göreme, however, the implications of the culture of competing to remain
equal is seen to be two-fold: on the one hand, attempts at innovation that fail, or
might fail, are ridiculed; on the other hand, successful innovation is emulated as
the ones who are behind attempt to catch up with the innovators. Though initially
criticised, the early innovators gradually became the leaders in a trend that many
other villagers would follow. Now that tourism has been proven a prosperous
venture, there is even a sense of heroism attached to having initiated the trend,
with many local men making claims to being among the first to have got involved:

> I opened the first restaurant. A few people saw that I was making money, so
> they also wanted to do something. Then they earned money, so others
> started. It started like that and so it carried on. They followed each other of
> course. If I hadn't done it, no one would have done it...One or two years
> after I opened the restaurant, I bought a car. Others in the village saw this,
> so they also started opening *pansiyons* and restaurants...In Göreme, twenty
> years ago there was one *pansiyon* – built from an old house. Then others
> did it, then they added more rooms, and it grew.
>
> (Abbas [*trans.*])

In this way, ideas concerning tourism business gradually caught on in Göreme, with more and more villagers emulating the visible success of men like Abbas. Abbas opened his restaurant and a few people converted some rooms in their cave-houses into *pansiyon* rooms for tourists to stay in. A few others had the idea of selling carpets to tourists. During those early years, Haci's Place and the few small *pansiyons*, together with the first carpet businesses, began the process of working together for commission, whereby the *pansiyon* and restaurant owners would get a small cut of the profit if they recommended their customers to buy carpets. There started the process through which the tourism businesses in Göreme became an entangled network of friends and partnerships, while at the same time being the source of bitter competition and rivalry among the villagers.

The capital used to build the businesses came from a variety of sources. On this, Tuncay, the carpet seller quoted above, told me:

> The people are just doing their own job and their own businesses with their own possibilities, rather than getting credit from the bank or things like that – some do but very few.

It was noted earlier that the Turkish government operates on a series of five-yearly development plans. These plans put in place certain financial incentives and credit schemes, which are usually most easily accessed by the larger, more formal, business organisations (Tosun 1998). In Göreme, villagers have generally neither had the know-how nor the financial capital needed to access such credit schemes. Most businesses, therefore, started small and basic and grew over time, with the gradual accumulation of profit fed back into building up the business. A few men sold the trucks they owned and started businesses, while others received money from relatives working in Germany. It is important to note that villagers have not sold their land in order to raise the capital for tourism business. Hence, tourism development has encouraged the labour within each Göreme household to become further split along gender lines: the women continue to work the fields, while the men bring in an income from tourism business.

While the above-mentioned government incentive policies introduced in the 1980s did not affect Göreme directly in terms of large-scale companies moving in, they probably did serve to boost Göreme's tourism growth indirectly by leading many of the independent tourists visiting Cappadocia away from Ürgüp in search of a new 'backpacker' destination. Villagers view the late 1980s as being the heyday of Göreme's tourism. Tourists seemed to flock to Göreme then and so there were always plenty of tourists to share around. The Gulf War is frequently cited as the cause of a dramatic downturn in tourist numbers at the start of the 1990s, and as such woke villagers to the precarious nature of tourism business. The Gulf War certainly gave the villagers a shock, and even during my fieldwork time going into the late 1990s, besides the few weeks during late July and August, there were fearful complaints sounded throughout the village that tourists were no longer going there. Tourists at first, during the

Plate 5.1 A new tour agency is built.

main growth period, appeared as a great and possibly unlimited chance befalling the village, but they later become entwined with the 'image of limited good' (Foster 1965) and this, as we shall go on to see, has fuelled the often heated competition among tourism entrepreneurs.

The early period during the 1980s, then, is often looked back on today with an air of nostalgia, especially among those who were the first to enter into tourism business. Abbas and his contemporaries seem increasingly disillusioned these days with the processes of rapid change going on around them. While once they were the main perpetrators of that change, they now seem to have been left behind by the younger men who have followed in their footsteps. Reminiscing about the early days, Abbas told me:

> In those days we always enjoyed ourselves, making jokes and so on. I used to close the restaurant, lock the door at the end of the evening, and go to Ürgüp with the tourists – until four in the morning. When I remember those early days my stomach churns with feeling.
>
> [*trans.*]

Indeed, the simpler and more informal character of those earlier businesses placed the village men in more of a position of 'hosts' to their tourist 'guests'. With fewer restaurants and tour agencies in the village at that time, *pansiyons* were the main centres of tourists' entertainment and the *pansiyon* owners were the main providers of that entertainment: serving meals, guiding on walks and trips in their cars, and singing Turkish folk songs when the tourists gathered in the evenings.

Today, while such services still prevail and tourists can still be entertained within many of the *pansiyons*, businesses in the villages have become more specialised. With the significant number of prominent tourist restaurants and bars along the main street of the village, tourists are now more likely to go out of their *pansiyon* for their evening entertainment. The guiding of tourists has also become more formalised as the number of tour agencies in the village has increased: in 1994 there were four tour agencies, in 1995 there were twelve, and in 1996 there were fifteen. As the businesses have become more specialised, so they have tended to become more business-oriented in the way they deal with tourists. This was expressed by one entrepreneur, who reflected that:

> Now it's only for money. Before we earned money from the tourists during the day, and then we went out with them in the evening and spent that money on them, all together. Now we've become cleverer, before we were a bit crazy.

A gradual increase in the effectiveness of laws concerning licensing and taxation has pushed Göreme's tourism businesses along the continuum from what has been termed the 'informal' sector towards the 'formal' sector economy (Hart 1973).[5] This shift has made it more difficult to run multiple services from individual businesses. Because tour agencies are now under the control of the national controlling body, TURSAB, which legally enforces the necessity of full insurance and licensing with certified guides, *pansiyon* owners therefore take legal risks nowadays if they take their guests out for a sightseeing trip in their car. Moreover, with the intense competition between the various businesses, tour agents in the village who do have the correct licence may be quick to inform the authorities of any illegal competition. These two themes, of an increase in competition and an increase in the 'formal' character of businesses, have characterised the growth and change of tourism in Göreme. Moreover, they strongly influence villagers' relations with tourists because they affect the ability of entrepreneurs to play at being 'hosts' to their 'guests'.

Tourism business and employment today

With more and more men trying their entrepreneurial hand in one or various tourism services, the rather haphazard process of opening new and copying existing businesses continued on into the late 1990s. These businesses are aimed almost entirely at foreign backpacking tourists. Some bars and restaurants are also frequented by local men, but otherwise it is only the grocery stores that are frequented by the villagers. A couple of small, more exclusive hotels, owned and run by non-Göremeli entrepreneurs, aim their services at the smaller of the bus tour groups. One of these is specifically aimed at a Japanese market. A few 'motels', housed in new buildings and situated just outside the village, together with another larger but not so up-market hotel that is owned partly by the municipality and is part of a Turkish hotel chain, accommodate the minority of Turkish tourists who stay in the village.

Besides the links that these hotels and motels have with national tour operators, there are few formal links between the tourism services in the village and tour operators or hotels outside the region. Most of the custom is thus 'caught' off the street, though more recently some bookings can be made with a few of the more up-market *pansiyons* and boutique hotels, as well as with the hot-air ballooning company, via their advertisements on the Internet. Many *pansiyon* owners do, nonetheless, have informal links with other accommodation establishments around the backpacker circuit, particularly in Istanbul and along the south and west coasts, whereby they send customers to each other. Personal networks seem to play a more important role in the ability of villager entrepreneurs to establish organisational links than any formal business networks.

A village-based organisation involved in tourism and the promotion of Göreme is the 'Tourism Co-operative'. This was founded in 1986 and, with over 150 village households as members, it allows those villagers who have not become entrepreneurs in their own right to have a part in Göreme's tourism business. The villagers in general seem proud of the work their co-operative does. The organisation produces posters and leaflets of Göreme, which it has distributed at tourism fairs in other parts of Turkey, and it has a general store in the centre of the village, a tea house and snooker hall (which opens for villagers on winter evenings), and a souvenir and drinks shop at the entrance of the Göreme Open-Air Museum.

The Tourism Co-operative could be described as a large village partnership system. Such systems (*ortaklık*) are rife in Göreme, as many of the privately owned tourism businesses are also owned between two or three partners.[6] A few people can raise the capital to start up a new tour agency more easily than one man on his own can, and such a system also shares out the risk involved. Many men have partnerships in a few different businesses, thereby spreading out their risk and also increasing the chance of earning a reasonably steady income.

The income obtained from these 'semi-formal' businesses is extremely difficult to gauge because information of this sort from entrepreneurs is inevitably clouded by issues to do with taxation and also competition and jealousy, and also by the *ad hoc* way in which money, employees, commission and so on are handled on a day-to-day basis.[7] Certainly the carpet shops have an enormous turnover of capital, and some of these businesses, together with the more successful *pansiyons* and restaurants, are now said to be worth hundreds of thousands of US dollars. One of the larger, more successful restaurants in central Göreme reportedly has annual takings of around US$80,000. Entrepreneurs who lease the buildings for their *pansiyon* businesses, on the other hand, make a much more meagre income. The yearly rental for a *pansiyon* averages around US$8,000–10,000, but with tourists being charged only US$5 each per night, operators might not earn back much more than the rental. They therefore aim to make their money by earning commission from selling other things to their guests, such as tours, carpets and hot-air balloon flights. They can also usually eat for free in the restaurant to which they most often send their guests.

An illustration of the figures concerning employers and employees from inside and outside the village is provided in a breakdown of the businesses carried out by Bezmen (1996).[8] Bezmen shows that out of a total of 188 tourism businesses in Göreme at that time, 161 were operated by local men. Out of the total of 166 people employed to work in these businesses, however, only 50 were local villagers. These figures illustrate clearly that business in the village is largely in the hands and control of local villagers. As well as being significant in view of the villagers' relationships with tourists (as discussed in the next chapters), this point bears strongly on the villagers' lives more generally. Because villagers are the owners of their businesses, they are their own bosses and so, in comparison to their employees, are largely free to do what they want when they want. This was shown to be particularly significant during harvest times in the village, when I observed that employees had to take unpaid time off in order to go to the fields to help their wives in this busiest time in the agricultural calendar. Entrepreneurs such as Abbas, in contrast, had their employees on hand to enable them to go off and help their wives whenever they needed to. Similarly, these men are free to go out fishing, hunting, playing cards and chatting with friends whenever they wish.

It was shown in the previous chapter that, because of the gender segregation in Göreme society, tourism work is considered a man's activity: it is inappropriate for women to be in the view of unrelated men, and thus to work in the 'public' sphere. Generally, the village women are not employed nor present in tourism businesses, and they have very little contact with tourists. A few village women are, however, gaining employment as cleaners in some of the more successful *pansiyons* and hotels. Although they usually receive low wages, this income is probably the only way in which village women can independently benefit financially from the tourism establishments in the village. Indeed, Scott's study of women's employment in tourism in Northern Cyprus comes to similar conclusions. Contrary to studies that have concluded that small-scale rural tourism can create business confidence and enhance the position of rural women,[9] Scott concludes that:

> women play a marginal role in small-scale family-run hotel businesses, and on the whole fare better in larger establishments with a more formal and bureaucratic employment structure.
>
> (Scott 1997: 61)

Just a few 'family *pansiyons*' still exist in Göreme wherein wives are involved in cooking and serving tourists. Most of the *pansiyons* that started off as family-run operations are now run mostly by the men, as their wives have been gradually separated from the 'tourism realm'. The financial benefits women receive from tourism generally come indirectly through their husbands and sons.

Göreme women's direct economic gain from tourism, if any, is largely on an informal basis and earned through chance meetings with tourists in the back residential areas of the village.[10] Along with the elderly, women in Göreme have come to represent the 'traditional' to tourists, because they mostly continue with the 'traditional' village activities of farming and food production, and they have not in tourists' eyes been tainted by the 'commercialism' of tourism. The contact they do have is shrouded in this air of 'tradition', and some women have come to make economic use of these representations by inviting tourists to view their 'traditional cave-homes' and then attempting to sell them self-made handicrafts. This 'traditional' representation of Turkish women has now been appropriated by the carpet shops, which have taken to hiring 'traditionally' dressed women to sit and weave carpets all day as a tourist 'show' in front of their shops. Rather than being Göreme women, however, these are women employed usually from mountain villages in the Adana region of Turkey. Again, one carpet shop began the idea originally and then all the others followed, even using the connections made by the first shop in order to contact other potentially willing carpet-weavers in the Adana area. Able to earn far more money from this tourism venture that they would from cotton-picking in their home villages, these women come and stay in Göreme for the summer season, chaperoned by a male relative for their protection. For Göreme women, working outside in the central area of the village in this way would be considered dishonourable.

In their role as entrepreneurs, village men are having to relate to people in new ways, and to do so, it seems, they are bringing their paternalistic attitudes and behaviours directly from the household. The workers, who are mostly young men employed to do the more menial tasks of cleaning and waiting in restaurants and *pansiyons*, come largely from other parts of Turkey and are often treated much like the *gelin* (incoming bride). While they are employed, they are expected to show absolute loyalty to their employer and the business they work for. They work extremely long hours and receive poor wages in return for their chance to 'learn the trade'. As we heard from Hüseyin in the quotes above, 'when you are in a *pansiyon*, you must do everything', including greeting the tourists, cooking, washing the showers and so on. Unlike *gelin*, though, most of these boys and young men can quite easily escape from their situation, and turnover is consequently fairly high.

A main bonus of working in or running a *pansiyon*, beside the chance to meet tourist women, is the opportunity it brings to earn commission, either by taking tourists to carpet shops or by recommending a particular tour agency for a day tour.[11] Touts in the street are generally considered inappropriate in Göreme. There used to be regular brawls at the bus station, as there still are in other Turkish towns, when a bus loaded with new backpackers arrived. *Pansiyon* owners, or people employed by *pansiyons*, would fight over the tourists. To avoid this trouble, and to prevent Göreme's tourism from developing a bad name, the Accommodation Office was set up to display advertisements of all the village's *pansiyons*. Tourists are now invited to browse and choose for themselves among the sixty or so establishments on offer.

So, while everyone does play the commission game, they do so from a position of employment, so that particular *pansiyons* have reciprocal links with particular tour agencies and they send customers to each other. Commission on carpet shopping is more of an individual system and most employees, as well as entrepreneurs and business owners, try to take tourists to their friends' carpet shop. Tourists themselves are also often employed on a commission basis as 'catchers' to draw in custom. It is thought tourists might believe a fellow tourist's sales pitch more than a Turk's. As such, foreigners are frequently used as pawns in the competition between the various businesses.

Volcanic eruptions: competition, fights and gossip

The competition that capitalism and entrepreneurship have brought about seems to dominate social relations in the village. Concurrently, however, the necessity and the ability to co-operate hold their ground and a sense of community prevails. This is because competition always operates within the context of the personal histories and relationships that exist among any community. At times, though, particularly during the summer season when business activity and competition is at its most intense, Göreme feels rather like a rumbling volcano that might erupt at any moment. As the season wears on and tempers wear thin, eruptions do occur. The following is just one example.

It was getting towards the end of the busiest season, and over the loud-speakers around the town the mayor called all of the owners of tourism businesses to a meeting in the central café. His main objective in calling the meeting was to ask the entrepreneurs to donate money to the Göreme Middle School for a new photocopying machine. Many of the village businessmen attended the meeting. After the mayor's business was over, one of the men handed out copies of a notice that had apparently been placed in each of the rooms of one of Göreme's more up-market hotels. The hotel is owned and run, together with an adjoining carpet shop, by a partnership comprising a non-Göremeli Turk and a woman from New Zealand. They had placed a notice in the rooms of their hotel warning their guests that some of the Turks they encounter in the village may be out to rip them off, even though they appear to be friendly; that many of the villagers you meet are only after taking you to a carpet shop where they will receive at least 35 per cent commission on everything you buy.

A guide accompanying a small tour group had found and been angered by the notice, and had taken one to show some men he was friends with in the village. One of those men had photocopied it to hand out to everyone at the meeting. The men attending the meeting were unsurprisingly outraged: How dare anyone dirty their reputation and honour in this way? How dare anyone jeopardise their relationships with tourists in this way? 'It is completely wrong', they said. 'What will it mean if tourists think we are just smiling at them for money?' Threats began to fly around freely:

They just do whatever they like and walk around with their noses in the air, completely comfortable with what they have done – *olmas* [impossible]! But now they will have to be careful. If they are driving their car out of Göreme one day, they will be stopped and beaten – just you see. Or some people might burn their carpet shop. Really, people are talking like that now. The Göreme people are very angry.[12]

In the days following the meeting there were various attacks on the hotel business: their advertisement signs were vandalised; tourists on their way to the hotel were intercepted and told to stay elsewhere; the owners received messages warning them to get out of Göreme. A few days later another meeting was called in order to hold a discussion between the villagers and the Turkish partner of the hotel, this time in the presence of the mayor and a tourism ministry official. The villagers jeered and shouted at the hotel owner, and they jeered and shouted at the mayor:

Why do you let outsiders come and work in Göreme? Why do you let them? It is your fault too – you let them come here and open their businesses, and they take Göremeli people's money, and then they spoil [*şimar* = get too comfortable or too big for their boots].

At face value, this whole incident seemed like a fairly straightforward case of competition between insiders and outsiders, and one in which perhaps the outsiders were using particularly nasty 'outsider' tactics. Underneath the surface of this story, however, runs a more individual vendetta between the business in question and another owned by a villager. There was a more personal history between those involved, the full extent of which not even the other villagers at the meeting were aware. At the second meeting, then, the hotel owner explained the full story so that the villagers were able to understand that the note in the hotel rooms was in fact a reaction, albeit a rather over-zealous and offensive one, to a situation fuelled initially by the one particular opposing villager. In the end, the village men at the tea house agreed to accept the hotel owner's apology as long as he promised to get rid of the offending notes in the hotel's rooms.

This series of events encapsulates many of the issues involved in the competitive and troublesome atmosphere that surrounds Göreme's tourism business today. It conveys something of the villagers' attitudes towards outsiders running businesses in their town; it shows the ways in which they both react and cope when riled; it demonstrates how precarious the villagers feel their businesses and their relations with the tourists are, and which aspects of tourism particularly affect their honour, both in their own eyes and in relation to tourists. These issues strike not only at the centre of the villagers' chances of future wealth from tourism, but also at the centre of their individual and village pride. Battling against stories in the media of tourists being 'hassled' and 'ripped off' by conniving moustachioed men across the band of Islamic Mediterranean and Middle Eastern countries,[13] villagers are constantly having to negotiate and recreate the impressions that tourists have of them. Thus competition between the tourism businesses involves a dynamic relationship not only among the businessmen themselves, but also between those businessmen and the tourists they are serving.

Causes of conflict

The point that tourists' custom in Göreme must generally be 'caught' off the street heightens the open and visible nature of the competition there. While some tourists follow up recommendations from their guidebook or from other travellers concerning which *pansiyon*, restaurant or bar to go to, others go from business to business in an attempt to get the cheapest prices possible. Furthermore, tourists' bargaining for the cheapest price also encourages competition, because entrepreneurs are drawn in by these 'budget' tourists' bargaining tactics in order to catch their custom, offering services at very low prices for the sake of taking custom for themselves and keeping that custom from competitors' businesses.

The previously discussed 'culture of equality' in Göreme means that the villagers are constantly competing to remain equal. This paradox in small communities has been discussed at length by Bailey (1969, 1971), who points out that:

> People remain equal because each one believes that every other one is trying to better him, and in his efforts to protect himself, he makes sure that no one ever gets beyond the level of approved mediocrity. Equality, in communities like these, is in fact the product of everyone's belief that everyone else is striving to be more than equal.
>
> (Bailey 1971: 19–20)

In practice, the villagers do not of course remain equal in their profits from tourism. Some have more business acumen or a knack of understanding what tourists want, while others have simply been lucky.

This 'culture of equality' is manifested in a variety of elements of behaviour and superstition. The 'evil eye', for example, is the envious eye and may cause harm to whoever it is directed at. Villagers pin small blue-eye beads on babies and girls in order to ward off any evil or envy that might afflict them because of their beauty.[14] The tendency to tread down anyone who is in a position to show off extends also to the possessions an individual may have. If a new acquisition is shown to friends too obviously, it is criticised in order to prevent the individual from becoming too 'big' in the presence of others. I usually found that any clothes I bought were 'bad quality' or I had 'paid too much' for them, and meat that I bought to cook for village friends was almost certainly going to be tough and of poor quality, even though it was not when we actually ate it. Villagers live by the saying, 'Bigger than you there is God' (*Sen den daha buyuk, Allah var*).

These equality mechanisms are strongly associated with the notion of jealousy (*kıskanç*). This is the reason given for the sometimes nasty business tactics used to break the visible success of competitors among Göreme's tourism services. Each entrepreneur is jealous when a competing business appears to be having more success at catching tourists than he is, and will resort to measures that will break that other's success even if those measures also damage his own profits. The frequent price wars that occur in Göreme during the summer months often reach a point where businesses are running at a loss, and this occurs precisely because of the strong desire in everyone that no one else become more successful than themselves. The idea that tourists are a 'limited good' adds to the pressure, showing that such aspects of Göreme's peasant culture (cf. Foster 1965) have continued through into their new lives living and working with tourism.

The increasing presence of non-villagers engaging in business practice in Göreme adds to the trouble. Increasing numbers of men from other parts of Turkey have come to the village for a chance of getting in on the act. As the village is the single most important element in a man's identity, those from a village just three kilometres from Göreme are considered to be 'foreign' (a *yabancı*, also meaning 'stranger'). A number of Kurdish men who have come to Göreme from the south-east of Turkey are sometimes treated by village men with an attitude that approaches xenophobia. In addition, a considerable number of foreigners from outside Turkey are starting to set up businesses. These incoming entrepreneurs frequently become quite successful, as they often bring with them hefty amounts of capital, and this financial capital is also usually accompanied by social and cultural capital. They may have solid connections with relevant sectors of the tourism industry in the region and elsewhere, and also more business acumen due to a higher level of education, training and experience. Moreover, foreigner entrepreneurs might know how to pull the right strings in the right places in order to get a good write-up in the tourist guidebooks, something that can make a big difference to the relative success of a business. Businesses run by non-Turks, or *turists*, frequently get the best write-ups in guidebooks, probably because their owners are able to talk the most convincingly to the writers of those guidebooks when they visit.

Indeed, while many Göremeli villagers still frequently claim that they 'don't know tourism', incomers have more confidence in their own knowledge concerning tourists' needs and wants. This refers back to the earlier discussion about innovation, because it is incomers, especially the *turists* who have become resident, who have introduced many ideas concerning the type and style of the tourism services in Göreme. Although it was Göremeli individuals who innovated the start of tourism, as tourism has grown many of the characteristics of the *pansiyons*, tours and restaurants were thought of by incomers and then continued to be emulated by villagers.

Rather than being valued for their input and ideas, however, these incomers become the object of jealousy and even contempt because of their frequent business success in the village. In conversation with one small group of non-Göremeli businessmen, for example, I was told:

> We've been to university. We think about what the tourists want. We ask what music they want to listen to and we play that. We give service. Then they [Göremeli men] get jealous, they come here, they fight. It's because we know tourism that the tourists come here. They [Göremeli men] don't know tourism. They don't ask what the tourists like. They have a cold atmosphere, no service, and then they come and beat us up. They are always jealous.

In villager discourse, on the other hand, outsiders come only to make money from tourism and therefore, without caring for their own or for Göreme's reputation with tourists, will 'steal tourists' money', thus tarnishing the experience and image of Göreme for tourism. One local tour agency owner described this in the following way:

> Compared to them [outsiders], Göremeli people are much better. In interaction with tourists Göremeli people are very good, very close, hospitable,...but outsiders came and now they've behaved a bit bad to tourists, and the tourists don't know – they suppose they are also Göremeli. In time, tourists won't come. And it will be because of the outsiders that they won't come. Göreme will be ruined.

The feeling among Göremeli men that their own individual standing in Göreme's tourism is under threat is translated here as posing a threat to Göreme's tourism in general. Outsiders are seen as not playing by village rules and, in return, any competition or trouble is usually directed more aggressively at foreigners. Certainly, if any fighting is occurring outsiders will usually be blamed.

However, to paint a picture in which Göremeli people and outsiders are each homogeneous groups and always working against each other would be oversimplistic. The reality of the situation is more complex, largely because of the histories of the relationships between particular individuals. Different individuals, whether they are Göremeli or 'foreigner', attain varying levels of respect among the villagers. The respect accorded generally goes by their perceived goodness of character, which in turn follows their previous behaviour and interactions within the village.

I have thus introduced some of the factors in the social and tourism make-up of the village that contribute to the presence of troubled competition. The 'type' of tourism present in Göreme, whereby tourists 'buy' the services within the village rather than having pre-booked them before they arrive, serves to promote an active 'face-to-face' competition between the many *pansiyons*, tour agencies, restaurants and shops. In addition, the tourists themselves heighten this competition in their overt quest for cheap services. An apparent 'culture of equality', together with the presence of 'an image of limited good' associated with peasant economic relations, also creates a situation wherein individuals will resort to often nasty tactics to prevent others from having more of the 'good' than themselves, particularly if those others are non-Göremeli.

Competition processes and tactics

If we return to the illustration above of the incident surrounding the letters placed in the hotel rooms, then we can begin to see some of the tactics used in business competition in Göreme. The letters contained what was, according to villagers, a lie that was intended to deceive the tourists and lead them to believe that they could not trust the villagers 'out there', and that they should only trust the places recommended by the hotel owners. This sort of lying to and deception of tourists is one key tactic employed in the competition game. A culture of deception and lies is prevalent in the village, in any case, and is in many ways inseparable from the prevalence of gossip associated with the egalitarianism and 'event richness' of small rural societies.[15]

This 'gossip culture' is also very much a part of the system of honour and shame discussed in the previous chapter. Gossip is an accepted form of social control in the village, used explicitly to check an individual's inappropriate conduct. Gossip is, as Bezmen noted in his discussion of Göreme social relations, a form of 'indirect confrontation' (Bezmen 1996: 216) and tends to be used without any particular regard for 'truth'. The game of deception invariably involves the skilful manipulation of the boundary between 'truth' and 'lies' in order to effect a particular consequence.

Frequently, then, tourists become unwitting pawns in the game of competition between the tourism businesses. Most of the time they are unaware that the game even exists. From the point of their arrival in the village, tourists are told different things about certain businesses depending on who they are talking to: that the *pansiyons* over there do not have water or electricity for most of the day; that they will be ripped off at particular carpet shops; that the other agencies do not have insurance for their tours, and so on. In the scenario beginning this section, the tourists were given the more extreme advice not to trust anyone at all in the village.

A further illustration concerns the company operating in the village that runs hot-air ballooning trips for tourists. The company is sent most of its customers from hotels and *pansiyons* in the surrounding area. The foreign owners of the company were wondering why they never received customers from a particular hotel, situated very close by. The hotel had many Japanese guests, and Japanese were usually regular customers of the ballooning trips because of warm recommendations in the Japanese-language Cappadocia guidebooks. A small amount of research had told the ballooning operators that the hotel in question was not sending its guests to another ballooning company in the region. It was not selling ballooning to its guests at all. Rather, the guests were being put off from ballooning by being told that it was dangerous. The balloon owners could not fathom the situation; in their view, the hotel owner only had things to gain by sending his guests ballooning, in particular his commission from the fee. In a discussion of the situation with Göremeli men who know the way the competition game is played, it was deduced that the owner of the hotel was discouraging his guests from ballooning partly because he was jealous of the visible success of the balloon company (even though there was no direct business competition involved), and partly so that the tourists' money was saved for carpet shopping instead. Although the hotel owner did not have any partnership in a carpet business, he would probably receive more commission from carpet sales than from balloon-trip sales.

The competition game can be complex, and unsuspecting tourists can sometimes be manoeuvred away from what would be desirable experiences. Believing much of what they are told, tourists become the board pieces, rather than being players themselves, in a complex game of manoeuvres similar to the game of backgammon.[16] Village men play backgammon with extreme skill and speed (almost always beating any tourist who may dare challenge them), and this is the same skill and speed, I observed, with which they play at gossip and deception within the village.

While the pressure this culture of gossip imposes on individuals is felt by both villagers and incomers, people from outside of the village are engaged in the gossip process in quite different ways from Göremeli people. Since they are outside of the 'moral community' (Bailey 1971), outsiders may have a sense that they are 'freer' than villagers because they are outside the complex web of honour and shame. Simultaneously, however, this 'freedom' is what renders outsiders potentially so dangerous, because their actions cannot be checked by gossip in the same way as with villagers. This might be why villagers often resort to violence when checking troublesome and over-zealous competition from outsiders.

Some tourists who stay for a while and find employment as casual help in the tourism services find themselves embroiled in the business competition shenanigans far more deeply than they ever intended. As outsiders, their understanding of the rules in the game of gossip, deception and lies is limited, and those who attempt to partake often get it wrong and make some serious blunders. While some tourists are employed simply as waiters/waitresses or to help out in *pansiyons*, increasing numbers of tourists are hired by the businesses to 'catch' other tourists. Although it is the Turks themselves who are hiring them, because they have to compete with the others around them doing so, the village men become very disturbed by the presence of so many working foreigners, particularly during the summer months when Göreme seems full of them. One day, for example, a Göremeli *pansiyon* owner told me angrily that he had just thrown a group of Australians out of his *pansiyon* because they had taken a tour with an agency other than the one he had recommended to them. The reason for his anger was that the tour the tourists had bought was sold to them by a fellow Australian working for that other agency:

The worst thing is that I'm Turkish, but they'll believe a stupid-looking guy from their own country before they'll believe me. That's the problem here. For example, those tourists listened to me telling about the tour, and then went out and listened to an Aussie and believed him that that agency's tours were better. They believed him. I'm from here, so I know better. That is what we're getting pissed off about. So I threw them out. I said: 'Get out!' So they went to stay in another *pansiyon*. And I know they got ripped off by that Aussie guy – he took them to buy carpets – and I'm going to tell them they got ripped off.

The reaction of this *pansiyon* owner is quite typical of the Göremeli entrepreneurs. Another man, a *pansiyon* and bar owner who had himself been involved in fights concerning foreign workers, told me in a similar vein:

You know, tourists always believe tourists better than us, that's the problem. For example, if a tourist comes in here now and sees you and me sitting here, they'll go straight to you and ask for information, not me, it's guaranteed. But I'm from here, I know everything. This is my place, so they should ask me, shouldn't they? But we are put in second place by this.

I asked if this was why the local men become angry about the foreign workers, because it puts the locals in second place. He answered, 'No, not just this. It is also because of business. A foreigner can attract all the tourists to one place. It's because of that.'

The hiring of tourists as 'tourist catchers' is seen as putting certain businesses at an unfair advantage. The disturbance this causes is evident in the quarrels and fights that take place. This fighting is sometimes among the villagers who hire tourists, and sometimes against the tourists themselves. Almost every week throughout the summer there is some sort of troublesome eruption, often with the *jandarma* called in, because so-and-so hit a 'tourist catcher' and the tourist's employer went to hit so-and-so; it is a never-ending drama. A partner in a carpet shop told me at the height of the summer season:

It's got too much now, it'll have to stop soon. You know, it's not fair, we're all making business equally and if you employ foreigners to work, they make it too messy.

Indeed, every now and then when the troublesome situation builds to a crescendo, someone becomes agitated enough to call on the authorities to check on the illegal workers in the village. Tourist workers are frequently informed on by business competitors, and sometimes the *jandarma* are alerted to do a general search throughout the tourism establishments in order to expel the illegal workers.

This is one example of how the villagers use the external authorities to their own advantage in competition. Another example of this tactic is where the representatives from the Ministry of Culture Preservation Office in Nevşehir have been called on to check certain businesses that may be defying the preservation laws of the village. Consequently, villagers have been taken to court for illegal building work they have undertaken, such as building into a cave in order to create another room in a *pansiyon*. Some cases were instigated by villagers telling on their fellow villagers; fellow villagers who are now also their fiercest competitors.

Despite the trouble it causes, entrepreneurs still continue to hire 'tourist catchers', largely because they are simply following everyone else. As I said above, the tourism businesses have developed through a pattern of imitation, with businesses constantly copying the practice, style and even décor of their competitors. The hiring of tourist workers is just one factor in imitation tactics, and is undertaken, like much of the business practice in Göreme, without a great deal of thought about whether it is actually 'good' business practice. Many tourists I spoke to said that they did not like to be served by fellow tourists when they were trying to see something of the 'real Turkey' in Göreme. I sometimes told this to tourism entrepreneurs, but my attempts at advising them were usually met with their reasoning that it must be good: there are Australians working in such-and-such restaurant and English girls working over there, 'so it is a good thing to do'.

In this way the businesses continue to almost blindly follow each other. When one restaurant decorates its exterior with a wooden terrace awning and cane tables and chairs, all the other restaurants along the street follow. When one travel agency adds a difference to the itinerary of its day tours, the other agencies follow. Copying then, is another tactic in business competition, tending to ensure that one's own business will at least remain equal with the others, even if it does not manage to achieve a higher level of success.[17] However, these copying tendencies also cancel out most entrepreneurs' enthusiasm for trying anything new. I heard a tour agency owner, for example, telling a guide who was complaining of being bored because the tours are all the same throughout the village:

> How can we do anything else? Then the other people will do the same, then they'll lower the prices and we'll end up losing money.

One strategy to curb the negative effects of competition is for villagers to work together to form partnerships and alliances. These alliances can take various forms. One form, as mentioned above, is where *pansiyon* owners either open or take up a partnership in a tour agency, so that the tourists from their *pansiyon* can be strongly recommended to go to that particular agency when they buy a day tour of the region. Some agencies have a few *pansiyon* owners as partners, some of whom may also be partners in a bar or restaurant or shop, so that a particular web is formed in terms of where and to whom tourists who enter the web are recommended to go. Constructing these webs of businesses therefore serves to trap tourists into one's own business network, thus broadening one's chances in the business competition, while simultaneously decreasing that competition. The webs are quite fluid: they change and mutate each season, and much talk occurs in autumn and winter planning partnerships for the next season.[18]

Another more informal version of this web process is the commission system. The dynamics of this system are extremely problematic and contribute to much of the trouble that occurs in the village, as commission alliances are constantly created, dropped and double-crossed. There are also heavy moral overtones associated with commission: one entrepreneur told me that 'commission is *haram para* [dirty money]' because it is stealing from the tourists. Indeed, commission does raise prices for tourists, as the price of a carpet is increased to include the payment to the person who took the tourists to a particular shop. Yet a strong belief exists among the entrepreneurs that they need this system in order to 'catch' customers.[19] Moreover, commission is a mechanism whereby all of those people who have not yet managed to be entrepreneurs in their own right, such as younger men, may profit from tourism. So despite the trouble and strife that it causes, it might also be argued that commission ensures the wider distribution of income from tourism.

Another alliance system is where associations have been formed between the *pansiyons* and the tour agencies in order to impose fixed prices. The Accommodation Association was formed primarily so that the above-mentioned Accommodation Office could be placed in the bus station with advertisements of all *pansiyons* and hotels, thus preventing jostles and fights with arriving tourists. The Accommodation Association meets at the beginning of each season and sets a minimum price for that year. The aim of this is to prevent any *pansiyon* giving way to tourists' bargaining to the point where one particular establishment has all the custom but no one has any income. The Tour Agency Association does the same, and each member signs an accord agreeing that any member who breaks the price code is punishable by a large fine. Such associations between small businesses in the tourism context have been formed elsewhere, such as in Thailand (Wahnschafft 1982) and Nepal (T. Kohn: personal communication). It is important to consider, though, the differences between those associations that are formal structures imposed on the businesses by government authorities, and those that are initiated by the entrepreneurs themselves. Wahnschafft (1982) noted that a taxi 'co-operative' with set tariffs was put in place by municipal authorities in Pattaya in order to protect tourists from taxi drivers overcharging. In Göreme, conversely, the associations were set up by the villagers in order to protect *themselves* from undercharging.

These alliances in Göreme thus work as a mechanisms by which the villagers protect themselves from tourists' bargaining and simultaneously from their own competitiveness. Villagers are attempting to avoid the situation where they might undercut each other 'to death'. A problem occurs, however, when, as each season progresses, pricing codes tend to be broken as individual busi-nessmen bow under the pressure of the bargaining from zealously 'cheap' tourists. The fixed pricing seems to be in conflict with the cultural importance in Turkey of negotiation taking place between buyers and sellers. Villagers simply cannot resist the temptation to enter into the habitual conduct of *pazarlık* (bargaining) over the services given. In addition, some entrepreneurs told me that villagers use the prices of their services as a way of arguing with each other, even when their argument started off being about something entirely different.[20]

There is without doubt a high level of tension between the opposing tenden-cies to compete and to co-operate among the villagers engaging in tourism business. Tourists walk around the village trying to find the cheapest services, and in an attempt to bargain they might lie to a tour-agency owner, telling him that another agency offered them the regional day tour for cheaper than they actually had. Tourists assume that because the agencies are competing against each other, they will offer the tour for less than the others. What they do not realise is that the agency owners are very likely to telephone around to check whether or not what the tourist told them is true. If they find it is true that the other agency has lowered its price, they will discuss or argue it out. If they find it is not true, then they will argue angrily with the tourist and tell them they do not need their 'crooked' custom.

Villagers who are competing against each other in business are likely at the same time to be close friends or relatives. As illustrated in the description of Abbas's life in the last chapter, these competitors/friends are always popping in and out of each other's businesses to chat, to drink tea, to engage in the more friendly competition of backgammon and so on. The relations between the village men competing in business are thus rather elastic: at times they stretch apart from each other, and then they bounce back, but they are constantly connected. In conversation with Tuncer (the carpet seller quoted above) about this point, I was told:

> I am still very good friends with my business rivals. We always visit each other and have tea and talk. But, on the other hand, tourism came and so you have to take a part of it, like a cake...You hurt that they break your business saying something that isn't true. It is really becoming very hard, being friends for many years, but it is also the life and you must live somehow, and some days you may disturb him and other days he may disturb you.

As with the system of gossip and deception, there are firm rules and codes concerning the ways that this tension between mutual competition and co-operation is played out among villagers. There was one incident during my fieldwork when a man who came from outside and set up business in Göreme was murdered by a villager, most likely because he had way overstretched the competition rules.

Consequences and experiences of tourism business

It would be easy in this representation of the Göreme villagers and their changing social relations through tourism to focus only on the fighting and individualist drive that tourism has drawn out in the character of the villagers. Indeed, Bezmen focused on this aspect of village life when he argued that:

> When the critical moment comes, people tend to behave in a way which reflects their suspicion of corporate organisation. The individualistic char-acter of village culture with its heroic and egalitarian overtones blocks understanding of mutual co-operation.

> (Bezmen 1996: 143)

Concurrently, however, the villagers *do* form these associations, alliances and part-nerships, and this is surely co-operation – or at least a desire for co-operation.

Indeed, the alliances formed between the *pansiyons* and tour agencies through the Accommodation and Agency Associations might be considered highly sophisticated, and co-operative, business practice. The tour agencies are still competing with each other on service and type of tour provided, but at the same time they work together to the extent that they often 'sell' tourists to each other in order to balance out the load in each agency's mini-bus each day. Moreover, the price-fixing undertaken by these associations serves to keep competition operating on the basis of the quality of service provided, while immediately placing more control in the hands of the entrepreneurs than in the hands of their customers, the tourists: the act of price-fixing removes the bargaining power from tourists.

This demonstrates that the relationships local men have with each other cannot be isolated from the villagers' relationships with the tourists. The part-nerships, webs and alliances formed between the businesses are very much a function of the villagers' business relationships with tourists and, indeed, serve to strengthen the position of the villagers in relation to tourists. The tourists, concurrently, often feel embroiled in a particular business 'web' and thereby restricted in their choice concerning which particular services they patronise in the village. We saw from one of the scenarios described above how a *pansiyon* owner was angered to the point of throwing a group of tourists out of his *pansiyon* when they bought a day tour from an agency other than the one he had recommended. Entrepreneurs have a fiercely possessive attitude towards tourists, saying, for example, 'They are my tourists and no one else can have them.' If the tourists do then go to another business that is in another 'web', they are behaving in a way that dishonours the relationship with their initial 'proprietor'. The villager they belonged to in the first place might then break off the relationship altogether by throwing them out and telling them that he does not need their custom.

The 'event-rich' character (Ardener 1989) of small societies like this could be said to play a large part in the reasons and the mechanics of the competition and trouble. Because the tourism in Göreme takes place around a fairly small, closed arena (the centre of the village), everyone knows everyone else's business: they can see with their own eyes how many tourists are in the other tour agencies' mini-buses each day, and one restaurant owner can see how busy the restaurants next door and across the street are. There is, in other words, a definite 'face-to-face' character among the tourism businesses, and this appears to intensify the competition between them.[21] There is thus a strong sense in Göreme that something is always happening and things changing. Besides the physical appearance of the village, which changes rapidly as each business emulates the other, there is constant trouble occurring as arguments flare up into fights that may even grow to involve the majority of entrepreneurs in the village, as with the eruption described earlier. The heavy and troubled moods that periodically hang over the village can thus most frequently be attributed to tourism in Göreme.

Each year, therefore, as the tourism season draws to a close, an increasing sense of *morah bozuk* (broken morale) looms. Villagers' moods change and their patience wears thin as the summer goes on. In the springtime they are busy planning and creating new partnerships and are optimistic concerning their co-operative alliances with each other, but by the time the autumn arrives they are not only tired of dealing with the tourists but are also fed up of their squabbles and fights with each other. Winter is the time when tourism is slow and there is the chance for villagers to spend time with each other once more. However, it seems that it is increasingly difficult for the 'community' to pull back together again. The carpet seller quoted above also told me:

> I wish that the people would come together periodically and talk about their problems, and I wish they would be more honest so that I would trust him and he would trust me as well. I wish that there would be trust all the time. Without talking, without communicating with each other, I am thinking bad about them and they are thinking bad as well, because you don't talk and you don't know what's happening.

A further consequence of the troubles is that the villagers are yet to achieve a sense that they are 'professional' in tourism practice, and also a sense that they are in real control of what they are doing in their new lives of dealing with tourism. The men are still very much in the process of trying to work out what tourism is, what they should be doing in it and what it is doing to them. This is invariably due to the villagers having to learn to relate with each other in new ways. Since tourism is fairly new in the village there is no tradition to fall back on when they are wondering what to do. They must simply muddle along, creating ideas as they go, and learning how to manage not only their new types of relationships with each other but also with tourists, incoming workers and the relevant government officials. This feeling of scrambling around in the unknown is only beginning to lift now. With the emergence of a new generation of tourism entrepreneurs, such as Hüseyin, the youngest of the group described above, a new confidence is beginning to dawn.

Finally, there is a sense in which the Göreme villagers are constantly trying to reclaim their honour and status in relation to the tourists and tourism in general, because they are aware that the various processes and tactics described above, such as gossip, deception, commission webs and the influx of outsiders, all put their relationship with tourists and tourism in jeopardy. Aware that their over-zealous competition tactics often lower the trust that tourists have in them, the village men are constantly concerned with how to heighten their individual reputation, and the reputation of Göreme itself. Perhaps they are right to be concerned because, as one tourist said to me while complaining about the annoyance of being 'hassled' by Göreme businessmen:

> Sorry, it is not as if they are all like that. But it's just that if you are sitting in the mountains with a beautiful view of the sunset and there are two mosquitoes buzzing around you, you can't help focusing on them!

The Göreme villagers are well aware that their image, and indeed their livelihood, is tainted and jeopardised by the troubles that so frequently occur. This point re-emphasises the necessity of considering the ways that the relations between the villagers and the tourists are affected by the villagers' relations with each other and vice versa, rather than viewing the two sets of relations as separate areas (as they are often viewed in tourism studies). Through negotiations of image, identity and power, relationships among and between villagers and tourists in Göreme are inextricably mixed. It is these negotiations in their relationships that will be discussed in the next chapter.

Notes

1 Useful comparative material is provided in other studies that have attempted to grasp the emic view of tourism development and change within societies. These include Berno (1999), Crick (1994), Ingram (1990) and Schoss (1995).
2 See, for example, Chambers and MacBeth (1992), Jamal and Getz (1995), Murphy (1985) and Wanhill (2000).
3 Elsewhere, but in a tourism context, Michaud (1991) draws a comparison between the formal and informal tourism sectors in Ladakh, India. He finds that the formal sector, which is made up largely of exogenous players, is relatively organised and 'connected', while the informal sector, comprised largely of people endogenous to Ladakh, is comparatively disorganised, insular and yet highly competitive.
4 See also Foster (1965) and Shanin (in Teodor 1971) for discussion of the suppression of innovation in peasant society.
5 A continuum is preferable to a dichotomy when considering the differences between the informal and formal economic sectors discussed by Hart (1973): most of Göreme's businesses seem to lie somewhere between the two. Picard (1996) describes a similar 'formalising' process occurring in the Balinese tourist centres of Kuta and Ubud, and he also notes how this affects the 'host–guest relationship' in these settings. Michaud (1991) provides an insightful discussion on the interaction between these two sectors of tourism business in Ladakh, India.
6 In an article by Delaney (1993) on 'authority and co-operation' in Turkish villages, the concepts of *ortak* (partnership) and *imece* (community project) are discussed in relation to the co-operative aspects of Turkish village life. In addition to the more

informal co-operation between women in their food-production and household work, as mentioned in the previous chapter, Delaney discusses here the more formal *ortak* system among the men. This is a sharing arrangement whereby heavy farming machinery or a mini-bus, for example, might be bought co-operatively between two or more men.

7 These points are also made by Crick (1994) regarding tourism businesses in Kandy, Sri Lanka, although I may have had the additional problem that it was considered inappropriate by entrepreneurs that I, as a woman, should have an interest in the economics of their businesses.

8 These figures appear in Bezmen's unpublished doctoral thesis on tourism and religion in Göreme. It is unclear, however, exactly how these figures were arrived at.

9 See Castelberg-Koulma (1991) and Garcia-Ramon, Canoves and Valdovinos (1995).

10 This informal nature of women's economic gain from tourism has also been noted in other tourism locations. See, for example, Cone (1995) and Swain (1993).

11 This process is similar to that described by Crick (1994) regarding the 'street guides' in Kandy, Sri Lanka, who try to 'catch' tourists in the streets in order to 'guide' them to a guest house or shop so as to earn commission on whatever is bought.

12 I did not attend these meetings as this would have been inappropriate for reasons of gender: the meeting was held in the tea house and only men were present. This is therefore a local man's reported version of what happened and of how he felt after the meeting.

13 See, for example, Bowman (1989, 1996).

14 Delaney (1991) discusses the connections between the evil eye, gender and cosmology.

15 See Bailey (1971), Elias and Scotson (1994) and Simmel (1955).

16 A game regularly played in the village, as throughout Turkey, in which each player is constantly looking ahead and calculating the possible consequences of his moves in terms of their effect on the moves of his opponent.

17 Bowman (1989) notes a similar process occurring along a row of souvenir shops in Jerusalem.

18 There is no strong connection between these business alliances and kinship, a point also noted by Bezmen (1996). Partnerships are as likely to be formed between two good friends as they are between two brothers. The ways in which these alliances and networks are formed thus relate to issues of social capital (Woolcock 1998) and trust (Misztal 1996).

19 Crick (1994) also provides an in-depth account of the problematic dynamics associated with the commission system in Kandy, Sri Lanka.

20 It would be wrong to assume that tourism is the cause of all of the trouble in the village. For example, Stirling (1965) noted feuds, violence and guns in the Turkish villages he studied in the 1950s. However, there is no doubt that the competition between tourism businesses has heightened the number and level of violent troubles in Göreme.

21 Here we are reminded of the principle illuminated by Goffman (1959, 1967) that the 'self' is only as it is in interaction with others: the presentation of the self is what constitutes the self.

6 Close encounters

Interactions between hosts and guests

It has been shown in the previous chapters that Göreme has become a meeting place for a variety of different people whose worlds are quite far apart from each other, and for whom Göreme means many different things. Yet there are also important points at which they meet. They are each aware, to some degree, of their own and each other's roles in the tourist encounter, and so they bring to the encounter some mutual elements of understanding and meaning. A fundamental part of that mutual understanding is that the Göreme villagers will show a certain level of accommodation and friendship to tourists. Indeed, 'Turkish hospitality' is conveyed as being a core traditional virtue of Turkish people in tourist representations, as with national identities throughout the Mediterranean region. On this, though, Herzfeld has rightly argued that the notion of hospitality is 'actively *constitutive*, rather than simply a *component*, of the stereotype of Mediterranean culture' (Herzfeld 1987: 86). Similarly, then, hospitality should be viewed as constitutive of the tourist–host encounter, rather than being simply a component of it.

There is a clear sense among Göremeli people that in their village they are 'hosts' to their tourist 'guests'. While objections have been raised regarding the use of 'hosts' and 'guests' for discussing tourism relations because of the sheer commercialism these terms disguise,[1] I will show in this chapter how the roles of 'host' and 'guest' themselves are used by the Göreme villagers in order to negotiate and determine their relationship with tourists. It will also become clear how these roles are used by the tourists in Göreme in order to intersect and reach beyond the primary tourist gaze, as discussed in Chapters 2 and 3. Furthermore, central to the ability for the roles of host and guest to be played out are the particular conditions under which the villagers' interactions with tourists occur. These conditions are, on the one hand, that Göreme villagers themselves own and operate the majority of the tourism businesses there. On the other hand, the majority of the tourists saying in Göreme are travelling independently of a package tour group. In order to situate the tourist interactions in Göreme further, it is useful to begin by recounting a tourist encounter, one of many such encounters, that I observed occurring within the package tour realm of cultural tourism in Cappadocia.

Packaged interaction

This encounter concerns a Göreme family who live just outside the village in a cave-house set, just below the top ledge, into the cliff of a valley that has now become one of Göreme's main panoramic viewpoints. Bus groups of tourists often stop on this ledge to gaze over the valley, and the guides of some of those groups, in an effort to provide an experience of the 'authentic' cave-life of Cappadocia, lead their tourists down the rock-cut steps to visit this cave-dwelling family. The family have grown to see these tourist visits as something of an easy money-earner, and keep a pile of souvenir items such as folk dolls, headscarves and lace items ready to sell to their captive audience. I knew this family well, and often visited them in their cave-home to eat and chat with the mother of the family. On one occasion, a group of thirty middle-aged American tourists were led into the house by their guide who was, in this case, a woman from Istanbul.

The guide brought all of her tourists into the house, and behaved in front of them with the old mother as if she was her long-lost friend, seemingly to create the idea that the visit was an 'authentic' one and that her particular tourists were lucky to have a guide who was so intimate with a local cave-dweller that they were able to join her in a visit to this cave-house. Throughout the visit the guide's performance conveyed the point that she was bringing her group to 'real' places. She was very ill at ease concerning my presence in the house, but was calmed to an extent on learning that I was an anthropologist rather than 'another tourist', and was keen to explain my presence to her group in order that the authenticity of their visit be established. She told the group explicitly that they were guests in this house, rather than tourists, and sat them in the living room telling them to feel at home.

With the old mother at her side as a 'visual aid', the guide then proceeded to deliver a long speech about the family and the house. She told of how they had dug out the cave-house from the rock; how they live in harmony with the landscape; how the mother works hard in the fields but always has plenty of time still to sew and embroider; how they live self-sufficiently and do not need money; how they use animal dung for fuel, and so on. Through her speech, the guide conveyed to the tourists a particular picture of the woman and her family: a picture of them as traditional cave-dwelling people who live in harmony with the landscape and with time and with each other.[2]

The tourists seemed pleased to have an anthropologist in their midst. I somehow added to the authenticity of the situation; perhaps the fact that I was there to study implied that there was something 'real' there to be studied. Some of the tourists asked me questions about the local way of life, and in answer I tried to include the mother, through translation, in the presentation being made of her and her way of life. The guide, however, was quick to cut both of us off. She herself made no attempt to include the mother in the presentation being made of her. She was keen only to place a traditional identity on the family in order to accord with the representations of Göreme that people found in promotional images and tourist representations in general. By the end of the visit the mother, whose house it was, had not played any part in the negotiation of what the tourists were to experience, nor of how the interaction took shape. After the group had gone she expressed annoyance and, when I translated to her some of what the guide had said about the family's life, said that the guide was a liar.

The discerning tourists likewise seemed to see the whole encounter as riddled with paradoxes and lies. One man attempted to protest against the 'animal dung for fuel' story by pointing to the television and electric heater in the room. Another tourist wondered if there were any 'cash crops' grown by the family. When I started to explain that grapes were the main marketable crops that the villagers sold for cash, the guide again interrupted with the idea that the people here have no need for cash. The same tourist then tried to establish what the family did regarding the payment of taxes, and he probably would not have minded if the mother had been able to say that her husband was a taxi driver and that her son owned a successful *pansiyon* for tourists in the village. What the tourists did mind was that they were experiencing an 'authenticity' that was blatantly staged, to use a term from MacCannell (1976), by their guide. In addition, there was the troubling paradox that they were thirty American tourists all trying to have an authentic experience at once.

The authentically social in Göreme village

We saw in Chapters 2 and 3 that, along with promotions of the physical landscape around Göreme for tourism, come images of cave-dwelling 'authentically social' people whose identities become attached to pre-modern time in tourists' imaginations. Even when forced these days to mention the almost demonic presence of mass tourism in the area, many travel promotions still rely on images of tradition existing in hidden corners in order to attract potential tourists, as did this article in *The Times*:

> The modern world, in the form of mass communications, has come to Cappadocia, a once remote plateau in central Turkey; but beyond the enclaves of hotels, tourist agencies and brash restaurants, the old ways persist.
>
> (Brook 1996: 17)

This suggests that there are two separate spaces in Cappadocia, each signifying something different for tourists: the tourist spaces with hotels and restaurants, and the non-tourist spaces where 'the old ways persist'. For many villagers, too, the older back streets or upper *mahalle* (residential quarters) are where the traditional social relations and moral values of the village are still largely intact, while in the central area the moral fabric of the village is evidently corrupted.

This dichotomy in Göreme between tourist space and traditional space is also strongly gendered, as discussed in Chapters 4 and 5. For tourists, the central area represents the place where they can meet and have fun with like-minded travellers in the backpacker cafés and bars, and it is also the area where the villagers they meet are men running tourism businesses. It is largely the women of the village, then, who represent the traditional in Göreme because, in their headscarves and on their donkey carts, they satisfy tourists' images of traditional Turkish life. The back residential streets of the village have thus come to represent for tourists something akin to a living museum, and many tourists spend much of their time wandering through these narrow winding streets looking for experiences of what they consider to be the traditional elements of Göreme life: cave-houses, donkeys and carts, and villagers going about their daily lives.

The tourists do not have a monopoly on the gaze here, however, since the villagers, who are on home territory and usually in larger groups, are undoubtedly also gazing back. In the afternoons groups of women sit chatting and sewing in shaded parts of the narrow streets at the 'back' of the village. Sometimes tourists walk by and gazes are exchanged. The ideas in Göreme surrounding the concept of *gezmek* (to walk around) play a part in the villagers' perspective on these encounters. As was explained in Chapter 4, in Göreme women cannot or should not *gezmek* because to do so would be to expose themselves to the gazes of men. In other words, those who are walking around are exposed to the gazes of those who are fixed in 'their' place, and that is precisely what is occurring when tourists wander through the back streets of Göreme. Frequently, as I sat in villagers' houses drinking tea with the women, they would look out onto the street below, and whenever tourists walked by they called '*turist, turist*' and mocked their clothing or their apparently ridiculous behaviour. We were always behind a window and the tourists had no idea they were being gazed upon, let alone mocked.

Villagers are also well aware of tourists' representations of them and their cave-dwelling identity. In turn, tourist representations and photography are likely to have added considerable influence to villagers' own images of their cave-dwelling lives. Some villagers told me stories of tourists coming and taking pictures that they then took away and printed in magazines in their home countries, displaying the apparent 'backwardness' of the Göreme people. Similarly, the woman in the cave-house scenario described earlier performs her cave-life for tourists by letting them come in and visit her house. The images that tourists have of her are thereby constituted through the performance and she is simultaneously made increasingly aware of those images.

This awareness does not necessarily come only from tourism, however. Television and outward migration have long enabled the Göreme villagers to see how others live, and they are fully aware of the 'backward' connotations associated with their caves. These connotations can rub off onto the villagers as shame, as illustrated in the following extract from my fieldnotes:

> In Esin's house, we have just been looking through the photographs that Esin took with the camera I gave her. When we came to one of her making bread, her mother said 'Who's that?' I continued the joke and replied 'I don't know.' She continued 'Oh, just a *köylü* [villager/peasant], look at her

making bread!' Then we came to a picture of the group of women eating. 'Who's that?' 'Oh, just peasants eating *manti* [a traditional dish] sitting on the floor. Look at them in their *yemeni* [headscarves].' 'Who says things like that?', I asked. 'People do, on TV they do,' she said, gesturing towards the TV.

Although there is an element of shame in the villagers' self-image as cave-dwelling peasants, they have learned to play up to this image in their relations with tourists, and they often invite tourists into their cave-homes to have a look around and to drink some tea. While this would seem to be not too dissimilar to the situation depicted above concerning the group of thirty Americans who visited a cave-home, it contrasts quite significantly in the way the actual encounter unfolds. First, for the tourists within the village such encounters are not so blatantly 'staged' or pre-planned for them, and so they are well-suited to the tourists' desires to have more 'real', serendipitous encounters with what they perceive as the traditional in Göreme. Second, with no guide acting as a buffer between the two parties, such events are arbitrary and open to negotiation by all involved, since they allow the villagers themselves to play with the role of host in the encounter, and thereby take some control over both their relationships with the tourists and the way they are viewed by those tourists.

The power in hospitality

Through direct interaction with tourists and the imposing of a host–guest relationship, villagers are able largely to negotiate relations of equality and respect with their passing guests. Göreme people take pride in their 'hospitable culture', and the concepts of *misafirperverlik* (hospitality) and *misafir* (guest) are central to villagers' discourses regarding themselves, their lives and tourism. Hospitality is connected with honour (Herzfeld 1987) and even nobility (Heal 1990), and guests to any Turkish village must be treated with the utmost respect and generosity. The issue of hospitality can nevertheless be a confusing one for both the hosts and the guests in touristic encounters. Hospitality is considered by local people to be an integral part of their traditional culture, and tourists usually receive that hospitality graciously, often discussing the great friendliness, helpfulness and hospitality in their experience in Turkey. Yet it is a misconception that hospitality should come easiest to the guest, as it is an exchange that always places the host in a position of control over the encounter.[3]

The villagers are extremely sensitive to the issue of respect, and if they perceive that the tourists are in any way disrespectful, then they will not interact with them. Villagers are adamant about the importance of judging each individual on their own merits, and they have developed a remarkable ability to make prompt judgements concerning tourists' characters.[4] I was frequently told by villagers that tourists are like the fingers on your hand: they are all fingers, but they are also all different, and hence some tourists are good and some are not so good. Villagers are quite critical of tourists who do not appear to consider their position as guest in a foreign country, and who, for example, walk around underdressed or behave in a rowdy fashion. If asked what they think about tourists, a typical answer from villagers is 'They are our guests and we like them. If they respect us, we respect them.' And so it is that only the tourists who behave in a respectful manner can enjoy close interaction with villagers, and this is the background against which the identities of both parties are negotiated in relation to each other.

Such encounters might not always run completely smoothly, however. When a woman invites wandering tourists in to look at her cave-house, she usually proceeds to present a pile of headscarves for her visitor to buy at rather inflated prices. Some tourists are disillusioned by this, since in their view the encounter instantly becomes a tourist event and no longer one of true hospitality. One tourist who had been disillusioned by such an encounter told me: 'This place slides between being authentically real and what's done for tourists.' On this, it is interesting to note the way monetary exchange was managed in the cave-house visited by group tours described above. Attempting to construct the encounter so as to appear as 'visit/hospitality' rather than as 'tourism/profit-making', guides were always careful to hide the payment that they made to the family for allowing their group to view the house. This then posed a problem, however, when it came to the end of the visit and the tourists were faced with the stall of handicrafts that the family were overtly marketing. Interestingly, the guide with the American group in the scenario above told her tourists that it had been *her* idea for the mother to have a few hand-made lace items and headscarves to sell, so that she would be able to benefit in a 'home-economy' way from the tourist visits. Determined to keep up the 'authentic' gloss on the family, the guide did not want her tourists to know that the cave-dwellers had entered into market relations with tourists from their own doing. As Heal has noted, specifically relating to tourism:

> The American usage 'hospitality industry' suggests an immediate paradox between generosity and the exploitation of the market place. For modern Western man hospitality is preponderantly a private form of behaviour, exercised as a matter of personal preference within a limited circle of friendship and connection.
>
> (Heal 1990: 1)

Since the tourists in Göreme generally come from a cultural background that construes a strong dichotomy between friendships and market-based relationships, in situations like the ones I have described here where hospitality turns into an economic event, they feel duped because the two are irreconcilable. For tourists, 'gift' and 'commodity' exchange are essentialised (see Carrier 1995a), so that friendliness/hospitality and economic relations are viewed as two very separate and opposing phenomena. Indeed, we are told of the dichotomy between friendship and economic exchange by *The Good Tourist Guide to*

Turkey, which warns tourists visiting Göreme that 'invitations to view the insides of houses should be seen as what they are: low-key commerce rather than simple friendliness' (Wood and House 1993: 253). For the villagers them-selves, on the other hand, economic transactions are often negotiated on a personal level and so the two can coexist. As one local entrepreneur told me, 'Turkish people really take pleasure from giving hospitality. Whether it's for money or not, it's in our culture.' Moreover, even if the women and families who invite tourists in to view their houses do regard the situation primarily as one in which they can make money, they are really quite open about the fact that they are merely making good use of tourist representations of them as being 'traditional' and 'authentic' by letting them have their desired experiences for a small fee: the price of a scarf must be nothing, after all, compared to that of the camera draped around the tourist's neck.

Indeed, photographic interactions are another area in which tourists are expected by villagers to play at being 'good guests' (Sant Cassia 1999) in *their* village. Göreme villagers are adamant that tourists should 'ask first' before taking a photograph of them, particularly the older villagers who are fearful because of the negative meaning attached to images in Islam. I observed many situations where the younger men of the village who 'know tourism' acted as gate-keepers in this regard. One such incident blew up into quite a fury when a woman from New Zealand was trying to photograph old women on their donkeys as they returned from the fields through the centre of the village. At the time, the tourist was in a carpet shop and she kept popping out to 'shoot' anything interesting passing by. As per usual, this annoyed the villagers she was trying to photograph and they conveyed their annoyance by turning away or waving a stick at the tourist. The carpet salesman, a young man from Göreme, asked the tourist to stop taking photographs because it was evidently disturbing the old women as they made their way home from the fields. The New Zealander, however, became defiant about her 'right' to take these photographs: she told me in conversation later that day:

> The traditional life is disappearing and so of course we want to take photos of it, and they should respect that...They should respect our culture too, that we want to take photos.[5]

This situation is illustrative of the Göremeli salesman's positioning of the tourist as a guest in his village. He is asking her to be a good guest, and in doing so is attempting to assert his position as host. However, the woman's rejection of his plea is also a rejection of her role of guest in relation to him as host. Rather, she asserts her identity as 'tourist' and assumes the tourist 'right' to collect and consume touristic images. Tourists know very well that they cannot be closet voyeurs, hidden behind their camera, when they travel and take photographs of 'natives', and so many might feel uncomfortable when they do so. However uncomfortable they are, though, their strong desire to collect and consume images of these 'others' frequently wins the battle. Being perhaps the epitome of the primary tourist gaze and the power in that gaze (Urry 1990), therefore, the act of photography can simultaneously be a key point at which the tensions present in tourist identity and the transgressions of the roles and relations of host and guest come to something of a head.

If we consider where local people may play with and determine their role of hosts to their guests, on the other hand, their relations with tourists become altogether more complex and less one-sided. It is precisely the positioning of hosts and guests that enables the Göreme villagers to have a significant say in determining their interactions with tourists. Moreover, for many tourists – those who accept their position as 'guest' or 'good tourist' – it seems fair that the women should make the most of their 'traditional' identity and directly benefit from the tourists' desire to interact with and photograph them. As Feifer argues, in reference to the 'post-tourist' introduced earlier, 'Resolutely "realistic", he [the tourist] cannot evade his condition of outsider. But, having embraced that condition, he can stop struggling against it' (Feifer 1985: 271). Thus, by placing the tourist in the position of *guest*, villagers are taking some control of and demanding some sort of levelling in their relationships with the tourists they let in to look at their lives. As 'guests', the tourists *have to* oblige their hosts by accepting the hand-crafted headscarf offered to them, and they *have to* pay the small fee that is asked for it. As 'guests', tourists have certain obligations placed on them, since the guest is obliged:

> to accept the customary parameters of his hosts' establishment, functioning as a passive recipient of goods and services defined by the latter as part of his hospitality.
>
> (Heal 1990: 192)

This point has important further implications for tourist interactions and hospitality in the tourism services realm of the village.

Hospitality in the tourism realm

The hospitality offered to tourists in the tourism realm is altogether easier for the tourists to deal with. Here, where it is accepted that services and experiences are paid for, interactions between 'hosts' and 'guests' and the positions of each player in those interactions are more clear-cut, and all tourists expect the tourism realm of the village to provide them with what they need in terms of accommodation, meals, transport and information. However, they also expect in their 'non-tourist' quests to have interactions even in the tourist realm that are consistent with the particular locality.

The *pansiyon* accommodation in Göreme fits perfectly with these expectations for various reasons. First, the *pansiyons* are small-scale tourism businesses and so allow the tourists to indulge in the idea that they are not participating in more 'typical' tourist activity. They also allow the tourists to meet with other like-minded backpackers to swap tales of their travels and to experience an important sense of community in their travelling. At the same time, since many are set in converted cave-homes, Göreme's *pansiyons* are suitably 'other' and consistent with the place, allowing tourists to engage in the fantasy that they are also, for a time, cave-people.

Moreover, because these small businesses are mostly owned and run by local men, they allow for close and unmediated contact between tourists and villagers. The services and interactions in the tourist realm of the village have not been perfectly set up and smoothed over, and so tourists can expect a rather tumultuous but friendly production of services. The expectancy that things will not go smoothly is all part of the adventure in Turkey, and the amicable terms in which services are conducted provide the close encounters that these tourists are seeking. As villagers are constantly saying to tourists, 'everything is possible in Turkey'. Tourists frequently comment on how their experiences in the village are enriched by such happenings as the mosque calling to prayer and the loudspeaker announcing village news and events from the municipality office. Every few days a truck goes around the streets billowing out insecticide spray, inciting comments, coughs and clicks of the camera from tourists. On most summer weekends the streets are alive with wedding parties or circumcision processions. There is always someone to talk to: if not some other tourists, then a waiter inviting you into a restaurant or a carpet salesman engaging you in a friendly exchange in order to seduce you into eventually buying a carpet.

While the tourist realm of the village is a place of work for local men and a chance to financially prosper from tourism, the men also see their role in this

Plate 6.1 'Turkish hospitality': a tea-house owner entertains tourists by playing his *saz*.

realm as one of host to the tourist guests in *their* village. In a Turkish village, private space is not necessarily equivalent to domestic or household space, but rather it is equivalent to the entire village. It was shown in Chapter 4 through the explanation of the varying levels of head-'cover' village women wear in different areas beyond the home, that 'private space' to 'public space' goes outwards in concentric circles from the home. The ultimate public space is outside the village boundary, where women should not go at all unless accompanied by a male relative. It follows that the village itself is construed as private space. Thus, anywhere in the village tourists are in the position of guests to their villager hosts and also have the imposition of being guests placed on them.

I observed this when, together with a British friend, I drove to some villages in the hills in the southern Aksaray district of Central Anatolia. The road would always lead us into the centre of the village where an open space was overlooked by the village tea house. When we stopped the car, we would immediately be descended on by crowds of men who came out of the tea house. We would then be ushered out of the car and into the tea house, where we were given tea and fresh bread fetched from the village oven where the women were busy baking. In no way would it have been possible for us to wander freely as tourists through their village. If we did express a desire to leave the tea house and see something else, such as an old mosque or a cave that they told us about, we would be accompanied by one or two men who were appointed to be our guides and hosts. We were smothered so thoroughly with hospitality that any sense of tourist freedom was removed from us.

According to Göreme villagers also, tourists are guests in *their* village and should be managed as such. As the predominant owners of the tourism businesses, the village men are largely able to manage tourists through regional day tours and, in the evenings, through entertainment by way of barbecues, full-moon parties and trips to the disco. Thus villager efforts to spend time with and entertain tourists are more than simple enjoyment, and they are also more than a cover-up for dollar signs in villagers' eyes. For they are also an expression of the villagers' desire, as hosts, to assert their own position of power and control over the visitors in their village.

This controlling aspect of the villager hospitality is nowhere asserted more strongly than in the tourist *pansiyons*. Since *pansiyons* usually retain the courtyard surrounded by high walls characteristic of the older cave-houses in the village, they provide perfect spaces where the 'meeting' between tourists and their village hosts can be played out away from the 'traditional' elements of village life. In other words, they represent a sort of 'free' zone, or *liminal* space,[6] for both tourists and villagers. Tourists use the *pansiyons* as a space to recharge between explorations of the valleys and their night-life activities in the village's bars and discos. They also use the space to relax in, meeting with fellow travellers, and to experience the more 'touristic' hospitality of their hosts. For many local men, *pansiyons* represent a free zone where they can drink beer, meet (and sometimes sleep with) tourist girls and generally hide from the watchful eye of the elders. Some men, for example, might wear shorts in *pansiyons*, but then change into more appropriate longer trousers or jeans before going out of the *pansiyon* to walk through the village.

Furthermore, *pansiyons* are spaces where tourists' behaviours and attitudes can most easily be controlled by villagers. *Pansiyon* owners are able to 'check' tourists' attitudes on their arrival in the village. They then proceed to endow them with hospitality, offering a free coffee or beer and inviting them on a sunset-viewing trip in the evening, followed by a *pansiyon* barbecue complete with Turkish music and dancing. They continue to control their visitors' stay by recommending tours and walks, as well as which restaurant, carpet shop and tour agency to patronise. The exchange of hospitality places guests in a position of obligation to their hosts so that the tourist placed in this role is obliged to follow up particular recommendations made by the host.

Tourists often, therefore, get caught in the web of relatives and friends of their key *pansiyon* host and proceed to be entertained by them and receive abundant invitations from them so that there is little time, or freedom, to do anything else. This was felt by a travel writer who visited Göreme and ended up writing a story precisely about the problem that she did not have time to write there because of all of the invitations and adventures to which she was forced to succumb. She wrote:

> Turkey is a hard place to work. I had no idea how seriously the Turkish people take their hospitality, nor how devilishly difficult it would make my life.

> (Holmes 1996)

Of course, these invitations and adventures all serve to individualise each tourist's experience in Göreme, and so while, as suggested in the previous chapter, tourists can feel restricted by their supposed allegiance to a particular man or group of men and their businesses, they generally enjoy the serendipitous nature of the interactions.

Restricting hospitality

It is the sense of restriction, however, and the confusion that arises when the exchange of hospitality obscures the relatively clear-cut quality of market-type relations created by paying for services, that sometimes leads tourists to rebuff the hospitality offered to them. Relations are easier to understand if they are centred clearly on a market idiom, and by not accepting their hosts' offers tourists stay removed from ties of obligation with their hosts. I saw many incidents where tourists refused offers of food, drink or help because they were unsure of the villagers' intentions regarding payment in the offer. A village *pansiyon* owner described such difficulties in the following way:

> I mean the culture is different. Like in Göreme, in our culture, if there are cigarettes on the table you just take one, without asking. But for them

[tourists] you can't, you must ask. And when we have food on the table we say 'Come and join us.' They say 'How much?' And they ask me how they can get to Avanos to the market, and I say 'I'm going to Avanos, I'm going there anyway, to the market, so I'll take you in my car.' But they say 'No, we'll take the bus', and they don't come. Then half an hour later I see them in the market in Avanos. They took the bus...But we understand them because they're travelling a lot and they get ripped off everywhere.

Concurrently, because of tourists' reactions, villagers are gradually being put off from their sensibilities of generosity and hospitality. They either find that their hospitality is rebuffed or, conversely, they get a sense that tourists are abusing their hospitality as a result of the quest for cheap or even free goods and services. Herzfeld also observed this in relation to 'hippy' tourists in Greece:

The wealthy tourists can at least be exploited. The 'hippies' on the other hand, take everything, but own nothing that can be taken from them. Their presence is somehow an abuse of the system, because it subverts the balance of reciprocity between foreign exploitation and local cunning that tourism of the grander sort has helped to create.

(Herzfeld 1987: 82–3)[7]

As we saw in Chapter 3, the quest for cheap or, even better, free services is one strategy by which the tourists who stay in Göreme open themselves up to serendipitous events and thereby individualise and strengthen their travel narratives. Villagers themselves have come to understand that the tourists' constant bargaining is part of their 'backpacker' ways rather than being a result of actual poverty. The villagers know that a large sum of money is required in the first place for any tourist's flight to Turkey, and they also see many tourists trying to bargain a US$5 room down to US$3 one day and the next day going and buying a US$500 carpet as a souvenir. As tourism plays an increasingly prominent part in the local economy, the local entrepreneurs are becoming increasingly disillusioned by the attitudes of the 'budget-traveller'. Constantly affronted by tourists' over-zealous bargaining, some villagers seem to despair at the 'cheap' tourists who come to the village: 'They just bargain, bargain, bargain, and then they complain about the service.'

One of the many illustrations of this was an incident where a tourist couple came into Abbas's agency to ask for train information for a route that would take them ten hours instead of three hours by the more direct bus route, but would save them US$5 off the bus fare. After they left the agency, Abbas exclaimed:

Some tourists are maniacs, really! What kind of tourists do we have here? Tourism is sightseeing and spending money, but with these tourists they don't want to spend any money! They even get free food from us, from our gardens. They go out into the valley and stay in a cave, and get up in the morning and munch their way through our gardens: apricots, pears, apples, they munch around like sheep – they eat everything!

Both tourist and villager narratives are filled with tales of how they themselves emerged as the righteous hero from being 'ripped off' by the other. The tourist narrative usually ends with the tourist managing to bargain the price of the item or service down to a remarkably low fee, or perhaps indignantly walking away and taking their custom elsewhere. Conversely, the villager narrative usually tells of the villager's hospitality being abused by tourists who try to outdo their 'host', and expect and take more than should be given. The narrative always ends, though, with the villager managing somehow to give the tourists their comeuppance, either by throwing them out or by taking the moral high-ground by offering them even more for free: 'Go on, take it. I don't need your money.'

Besides the stories told, I observed many instances of tourist–host clashes in this regard. An example occurred one evening as I sat with Abbas in the door of his agency watching the goings-on in the busy street outside. In front of the shop next door, a general store selling provisions and also foreign-language newspapers for tourists, we saw a tourist woman sitting on the kerb reading a French newspaper. She was hiding behind the newspaper stand so that she would not be noticed as she read the newspaper for free. Abbas called to get the attention of the shop's owner, who responded by wandering slowly over to the tourist and, standing above her, said 'Hello' to rouse her attention. She looked up and, clearly embarrassed, placed the newspaper back into the stand. The shop owner then took the bottle of water the tourist had under her arm, took the lid off and offered some water to his friend standing nearby. The men laughed and put the water bottle back into her hand. 'Sorry', she said and walked off. She was embarrassed, and had been put clearly in her place by her Göreme 'host'.

Such an effort to rebuke tourists for their cheek is related to the issue of tourist respect mentioned earlier. As 'good hosts', villagers make a point of being sensitive to the character of each tourist, and entrepreneurs in particular are coming to view each tourist nationality as different in this regard. For example, one villager told me that while Americans are good tourists because they are relaxed, fun and open, Europeans are too serious and Australians are abusers because they are too cheap. Indeed, the increasing numbers of Australians visiting Turkey are gaining a reputation among Turks as being the worst among the backpackers: 'Australians are *pis* [dirty]! They have no culture and no respect. They bargain over everything. They don't want to spend any money!' Villagers are thus quick to discriminate between those who are and those who are not 'good guests', and are becoming increasingly intolerant of those who are not.

Another element in the tourists' behaviour that disturbs the villagers' ability to play at 'host' is the tourists' desire to meet and have fun with each other.

Entrepreneurs frequently complain that tourism in the village has changed during the past few years because the tourists are increasingly gathering in their own groups and only show interest in interacting with each other. The increasing hordes of Australian backpackers, such as the group depicted in the tourist portraits in Chapter 3, are usually blamed for this. As one villager complained,

> the Australians are changing tourism here because they all get together – travelling, walking, drinking. They don't mix with us. This is our town, you know.

Another entrepreneur told me:

> Tourism used to be better before, because it was all European tourists and they had really nice times. We always had barbecues out here. It was a nice atmosphere, nice conversation. Now it's ruined in Göreme, because it's all Australians, who aren't interested.

'Tourism's going down here,' he kept saying. This apparent change in the ambience of the village's tourism is possibly linked to the trend discussed in the previous chapter in which the tourism businesses are becoming gradually more formalised in their structure and services. As the businesses become more specialised with the increasing emergence of restaurants, bars and discos in the centre of the village, tourists now leave the *pansiyons* in the evenings and party in the bars and discos rather than having to submit to the more complete hospitality of their *pansiyon* hosts as they did in the earlier days of Göreme's tourism.

There has also been an increase in the backpackers' tendency to gather in groups. In the late 1990s this tendency was manifested in a 'hop-on, hop-off' bus service, started by an Istanbul-based and part New Zealander-owned company, specifically for backpackers travelling around Turkey. Now new hordes of twenty to thirty backpackers arrive in the village every two days, all booking into the same *pansiyon* and descending on the bars and discos in the late evening, together in their large group. Even though their time and activities are not managed as such by an agency and a guide, as were the package group tour previously depicted, groups of tourists such as this inevitably interact with the places and peoples they visit in different ways from backpackers travelling individually or in pairs. What is internal to the group inevitably takes on more importance than anything external to the group, such as the village and villagers they are visiting, and their cultural self-confidence (Graburn 1983) may gather such strength from the group that they become openly abusive to those outside the group. The villagers thus feel further abuse of their efforts to be hospitable, and are led to assert control over the tourists in other ways. I observed an incident one evening in which a group of about ten Australian men were sitting in the free-entry disco and not buying any drinks. They were also overheard criticising and cursing Turks, and word of this was quick to get around. Within a few minutes a crowd of local men arrived and sat near to the Australians. Nothing was said, but their intimidation was enough that the Australians soon left the disco. More recently, and especially since the advent of the backpacker bus company, villagers have resorted to more violent means to deal with tourist abuse, and fights and even stabbings are not wholly uncommon during the night-life activities of Göreme's tourism.

This increase in violence, like the competitive violence between villagers discussed in the previous chapter, places villagers in a state of flux regarding their self-image in relation to tourists. This was illustrated to me in Abbas's agitation the day after a robbery had taken place in his brother's *pansiyon*. Tourists had drugged and stolen from fellow tourists in a dormitory room, and he said to me:

> In your interviews with tourists you should ask them what they think of us, and if they think wrongly you should explain the truth to them. Göreme people are honest and good people, they don't deceive tourists. That robbery that happened, I bet that tourists will think it was Göreme people who did it. But it was tourists!

Abbas continued from here, however, to talk about how tourism had changed the village, and in this he expressed sadness that there was no longer such hospitality in Göreme because tourism had ruined it: 'Now we just smile at tourists to get their money.' One local man told me:

> It is very important for Turks to offer hospitality to foreigners – they even fight over visitors – 'he's mine', 'no, he's mine' and so on. *Paşas* used to build big houses so that they could take the most guests. But in Göreme it's finished, it is not hospitality now. They used to see a tourist passing a field and offer him grapes but now it's finished.

There is a clear sense among villagers, then, that their hospitality is becoming eroded through their dealings with tourists. This erosion of hospitality is felt as a loss of an integral part of villagers' identity at a variety of levels. A similar process is noted by Zarkia (1996) in her discussion of host–guest relations on the Greek island of Skyros, and Herzfeld also makes this point in relation to the abuse by tourists of villager hospitality in Crete:

> They [abusive tourists] were not simply strangers to the village, but also guests in Crete and in Greece. Since their behaviour violated the rules of local hospitality, it also violated those of the larger entities...[it was an affront] to the reassertion of domestic, local, and national sovereignty – to control over the metaphorical 'home' at all these levels.

(Herzfeld 1987: 81)

In Göreme, too, just as 'home' – from individual household through village to Turkey as a whole – is layered, so hospitality is also layered, operating at all levels of the villagers' identity.

The central place of hospitality in villagers' narratives is indicative of the fact that this issue is at something of a crisis point in Göreme's tourism. Many jokes and parodies performed by villagers regarding tourist–host relations highlight the strains. One example was when a friend of Abbas picked me up in his car as I was walking back from the Göreme Open-Air Museum one day. He told me to tell Abbas, as a joke, that he had charged me 500,000 Turkish lire (about US$3) for the ride, so that Abbas would get angry at his 'ripping me off'. When I then ate at that same friend's restaurant (as my way of repaying him for the ride he had given me), he then told me to continue the joke by reporting to Abbas that the meal was not nice and that I had been over-charged. On another occasion, a village man got me to phone up one of his carpet salesman friends pretending that I was phoning from Australia to ask where the carpet was that I had bought last year but had never been sent. A lot of fun was had from the salesman's squirming on the telephone and his reporting to us later of the difficult situation he had apparently got himself into.

Further parodies are made of tourists and their behaviour in relation to villagers. I saw a particularly clever example being performed by a carpet salesman who sat in the bar next to his carpet shop and started chatting to a group of Australians who had just arrived in the village. The Turk's good command of English and dress of jeans and T-shirt allowed him to play at being a tourist himself. He proceeded to play the wise tourist who had been in the village for a while and therefore knew what was what. He told the Australians that if he were them, he would not stick around long because it was too expensive and everyone here would try to rip them off, especially at that carpet shop next door. The tourists looked thoroughly confused: his accent and demeanour were slightly 'off' for him to be one of them and so they suspected that he was a Turk, but why then was he telling them these things? He was in fact performing with such irony that he was managing to make a parody of both tourists and locals at the same time, thus highlighting the tensions between them.

All of these parodies were clear attempts to deal with some of the tension points felt by villagers at their own and each other's behaviour in relation to tourists, and in turn at what tourism was doing to them. They seem particularly pertinent when, as we have seen from this discussion, villagers' expressions of regret over the loss of hospitality through tourism may be translated as expressions of the loss of a sense of control over individual tourists, over the village and over the nation in the face of international tourism. Goody (1977) notes the cathartic value of joking and humour and their function in relation to the management of conflict and, as MacCannell remarks, 'Parody builds solidarity in the group that stages it and potentially raises the consciousness of an audience that it is the butt of it' (MacCannell 1992: 32).

Close encounters of a more 'real' kind

To recap, this chapter has focused on the interactions and negotiations between tourists and villagers in Göreme as they compare to the types of relations evoked by the package group tours that are largely controlled by external tour agencies and guides. By taking a comparative look at the two different styles and structures of tourism in and around Göreme as the starting point of the analysis, it has been shown that while both package group and backpacker tourists follow tourist representations of the 'authentically social', what does differ is the degree to which their experiences are pre-structured, and hence the degree and quality of their interactions with villagers.

The guide of the group tour worked to place a stereotypical 'cave-life' identity onto the family they visited. This requires the preservation of a static cultural identity, because the group tour situation allows for no negotiation of this identity to take place between the tourists and the villagers. It is, then, the design and conditions of the cultural package tour that perpetuate stereotypes concerning local culture, rather than necessarily the representations and the primary tourist gaze in themselves. The tour group situation was seen to be unsatisfactory for both tourists and villagers: villagers are unable to offer hospitality to tourists and thereby to level their relationships with them, and tourists are left to juggle with the apparent contradictions between the representation and the reality of what is being presented to them.

The independent tourists staying in Göreme, on the other hand, buy into local tourism services and are therefore able to individualise their experience by asserting their own role as guest. This is part of their 'non-tourist' discourse, since the guest is one who avoids 'tourist places', who stays for longer and who 'hangs around' having meaningful interactions with local people and places. When asked how they liked Göreme, for example, three young German tourists replied:

> We like it here more than the coast because the people here are very open and friendly, and there are lots of nice places to go hiking. It's really different because the people are so open here, you can feel closer to them, you know, you don't really feel like a tourist here because you can have closer contact with the people.

These words neatly echo the words of a Göremeli *pansiyon* owner when he was asked in an interview how he thought tourism was generally going in Göreme:

> We want to keep Göreme for backpackers because if we build more hotels then we will lose the backpackers, and we've been doing this business for thirteen or fourteen years already. We grew up with backpackers, and it is wonderful. You can talk with them, you can learn a lot from them. The package tour people...don't have time. We cannot talk with them, spend time with them, because they're all organised. They come one day and then

they go the next day…Backpackers stay here longer so you talk with them, have fun with them and you get to know their lives too.

The similarity in tourists' and villagers' experiences in turn highlights the point at which, despite difficulties and contentions, tourist and villager discourses and experiences do meet in Göreme. Göreme meets with tourist quests in that the encounters tourists are able to have with villagers, both in the front and the back realms, are sporadic and unprepared or mediated by guides. This generally satisfies the tourists' desires for 'unstaged' and serendipitous experiences. It also provides the chance for the local people to place their own demands on the situation. As the owners and managers of tourism in Göreme, the local men are in the position to assert a host–guest relationship and to thus have a certain level of control over their relations with tourists. So while providing adequate services and entertainment for the tourists, villagers are able to demand and determine in various ways that their relations and identities are negotiated in a context of equality and respect.

Concurrently, however, the tourists' position as guests in relation to villagers' hospitality, particularly within the *pansiyons*, can be confusing for both players. In Göreme, obligatory ties are usually placed on tourists quicker than they could have expected, and though these chances of close and friendly interaction with local people do meet with tourists' quests for serendipity, they can also, as suggested in the earlier quote, make things 'devilishly difficult'. Situations often arise in which tourists feel confused about offers of generosity and friendship in the tourism realm, or where they feel trapped and restricted by the obligatory ties created by their villager hosts. Similarly, the villagers increasingly feel that their hospitality is abused and eroded by tourists.

While this is the case, however, and while relations between backpackers and villagers in Göreme can often be problematic, an increasing number of tourists enjoy being 'caught' by the villagers' hospitality and friendship, and to individualise their experiences further stay longer-term in the village. Many of these are women who develop sexual relationships with village men. Having looked in this chapter at tourist–host interactions more generally, I will now go on in the next chapter to focus on these longer-term relationships.

Notes

1 See Bruner (1989), for example. Zarkia (1996) has also argued that, because tourism transforms the host–guest relationship into a commercial one, the power is transferred from the hosts to the tourists, since it is they who have the money.

2 Chambers describes a tour to an Iban longhouse in Borneo in a remarkably similar way, with the guide staging and mediating the visit to, as she described them, 'a truly unique people' who 'continue to live closely with nature' (Chambers 2000: 67).

3 See Berno (1999), Heal (1990), Herzfeld (1987) and Wood (1994) for discussion of the social exchange of hospitality.

4 Crick (1994) also notes that people of Kandy in Sri Lanka, and particularly the entrepreneurs working the 'informal sector' of tourism, are good judges of tourists'

qualities, nationalities and characters. Such judgement becomes crucial in the tourism context as the livelihood of entrepreneurs may depend on it.

5   The power in the 'photographic gaze' has been discussed by many theorists: for example, Bruner (1989), Crawshaw and Urry (1997), Foucault (1977), Sontag (1979, 1983) and Urry (1990). It has been argued by Bruner, for instance, that in tourism, photography 'isolates the native people from their larger social context' and, as such, it 'decontextualises, and is essentially conservative' (Bruner 1989: 441).

6   Following Turner's description of ritual (Turner 1969), which in turn is based on Van Gennep's earlier work of 1909 (Van Gennep 1960), this term refers to the process in ritual where the usual order of things is removed and sometimes reversed. In connection with tourist experience, the concept has been discussed in particular by Graburn (1983, 1989).

7   This is also a point made by Riley, who notes that the status-enhancing experiences of getting a lot for little cost puts 'budget-travellers' 'in a position to exploit the hospitality of locals' (Riley 1988: 321). An important point to add to their observations, however, is that locals may regain their power precisely by providing that hospitality.

7 Romantic developments

New and changing gender relations through tourism

We say eye-wash. European girls are washing the eyes of the men. They're uncovering their legs, showing their arms, and putting on lipstick. Turkish women, especially Göreme girls, they don't know – of course they know lipstick by now – but they don't use it. And of course we go to the fancy one, nice one, pretty one, open one. She can speak with me about herself and I can speak openly with her. Because she is free and I am free, but that one [the Turkish one] is not free.

This extract from an interview with a local man explains how the men in Göreme are being drawn into relationships with tourist women that contrast with the kinds of relations they have with local women. Placed against the context of local gender roles and relations in this way, this type of tourism relationship – the local men's relationships with tourist women – is presented almost as an inevitability, as an opportunity difficult to miss: 'she is free and I am free, but the Turkish one is not free.' This type of tourism relation is becoming extremely prominent in Göreme, with an ever-increasing number of short-term 'romances', as well as long-lasting relationships and marriages taking place between local men and tourist women. A triangular set of relations thus unfolds between tourist women, local men and local women, giving rise to many important issues concerning not only the interaction between global and local, but also the links between gender and power. The 'romantic developments' in the title of this chapter are two-fold. First, this refers to the growing presence of romance, or at least an ideal of romance, in the local setting through and because of these romantic liaisons.[1] Second, there is a development of tourism business taking place in Göreme that is generated specifically from these relationships.

Sex relations in the tourism context are embedded in the cross-cultural complexities of gender, sexuality and power (Bowman 1989, 1996; Hall 1992). As was seen in the previous chapter, the close level of interaction between villagers and tourists is an important factor in the villagers' experiences with tourists; that closeness allowing the villagers, in part at least, to redress power inequalities inherent in the tourist–host relationship by asserting their own control over tourists' activities and experiences. Sex relationships might be a further way in which these men can regain a sense of control over their tourist guests, a sense of control that is otherwise experienced as diminishing as the level of tourism continues to rise in *their* place. This is precisely the way that sex relations between tourist women and 'host' men are explained by Bowman in his assertion that ' "Fucking tourists" in Jerusalem in the eighties was…a means of imagining and acting out a power that, in fact, the merchants did not have', because it provided them with 'a field in which to play out scenarios of vengeance against foreigners who, in their eyes, oppressed them both economically and socially' (Bowman 1989: 79). Zinovieff (1991) paints a similar view of Greek men's sexual relationships with tourist women, arguing that the men's tricking, lying and sexually conquering tourist women is a way of symbolically counteracting ideas of the women's and the West's underlying superiority.

However, like many studies of tourist–host encounters, these accounts fail to provide a balanced view of tourist and host narratives and how they relate to each other. Rather, they tend towards an over-concentration on the purpose and strategy of the men involved in such relationships, while playing down the voices of the women. Moreover, attempts to describe and explain these relationships seem repeatedly to look for their structure and function, thus neglecting the possibility of excitement and attraction. In other words, these relationships, along with most other contexts of cross-cultural courtship and marriage (see Breger and Hill 1998), are usually presented in terms of their outcome and as a means to a particular end. They are seldom viewed, particularly where the men involved are concerned, as processes in which the 'anti-strategy' of emotion may play a part (Kohn 1998).[2] Moreover, these relationships are often conveyed through the idiom of male 'predator' and female 'victim', thereby reiterating the gender stereotypes of rational and strategic men versus emotional and weak women (Seidler 1987).

By contrast, I aim in this chapter to develop an understanding of both the reasons *and* the emotions evoked by these tourism relationships in Göreme. The discussion here is based on interviews and focus groups with both local men and tourist women involved in the relationships, and I also include the views of local women regarding this fairly new social development in the village.

Fun and romance

Tourism business is largely the domain of men in Göreme and, although the tourism realm is therefore the men's place of work, it also represents something of a free zone in which the men feel relatively free from many of the restrictions normally present in Turkish village life. It is in this arena that tourist women and local men first meet, and where the men find themselves to be both the victims of the tourist's 'eye-washing' presence, and the lucky inhabitants of a tourist 'paradise'. While male tourists are accepted and welcomed, newly arrived women generally receive a great deal more attention.

There is a belief among some village men that they are more handsome, more willing and better in sexual relations than men in the tourists' home environments; they deduce from this that foreign girls actually go to Göreme for sex. The answer one young *pansiyon* worker gave to me when I asked why many tourist women come and have relations with Göreme men was: 'Because we are handsome and young, you know, nice *tak tak*. We can do it twenty-four hours!' The fact that so many tourist women do have relations with the men is clearly enough to prove the men's sexual prowess and thus to heighten their sexual identity. With busloads of new arrivals every day, tourism in Göreme has produced a sense of paradise for local men. Some men even referred to the Koran in conversations about this topic, telling of where it says that in heaven there will be forty women around each man. Göreme is like that now, they said; 'It's raining girls here!'

Similarly, the charm that the Göreme men display to new arrivals clearly appeals to the tourist women's sense of their own attractiveness, and in doing so heightens their sense of their own sexual identity. A woman from the USA said of her experiences in Göreme:

> I don't get looked at at home, then I come here and I've got ten guys all looking up admiringly at me. If there is any girl here who says she doesn't like it, she's lying. Any girl who didn't make the most of it and have a good time here would be stupid.

This was also expressed by two women from New Zealand who told me that although they had heard that women are hassled a lot in Göreme, they had found no problems there, especially after Istanbul. They added 'you get chatted up here, but it's no big problem' and 'it's nice to get a bit of attention, I felt quite bubbly when I was first here'. Moreover, as the American woman suggested, some women who 'don't get looked at' because they may not satisfy standards of beauty or ideal weight, for example, at home, can find themselves being the object of much amorous attention from Turkish men. So, just as the tourist women reflect a positive self-image back onto the men regarding their sexual identity, the men enhance the women's positive image of themselves: 'They are so charming – they make you feel like a queen.' This, together with the financial and cultural powers usually associated with the tourist in relation to the local people in the tourist setting, serves to enhance, for the time that she is on holiday at least, the woman's own sense of personal and sexual power.

Furthermore, the sense of enchantment surrounding these meetings is strengthened by the context in which they take place. The women are in a magical land of fairy chimneys and caves, and the men are in the tourist realm where they are free to play and experiment with roles and identities. The liminal nature of both the women's and men's experiences in this tourist realm allows for and promotes a sense of romantic and sexual freedom that might be more restrained in their 'home' contexts. He is in his new paradise where uncovered and 'free' women are plentiful; she has arrived in an enchanting landscape where she is charmed by numerous attractive and attentive men.

Of course, the women are usually aware that the men must have a family life somewhere 'behind the scenes', and for some women this point feeds into their ideals of the exotic in their interactions with the 'local'.[3] Conversely, many women have no interest in anything other than the fun and play of the tourist realm: fixed in the 'holiday' mode, they prefer to ignore the potential complications of the background of the men they meet. The tourist realm of central Göreme, together with the backdrop of fairy chimneys and caves, thus provides a magical and bewitching context within which these liaisons take place.[4]

However, the interactions between village men and tourist women are not without problems. After receiving warnings from family and friends, special notes for women in the backpacker guidebooks, and being 'hassled' by men in Istanbul, some women then experience men's advances in Göreme as annoying. The term 'hassle' is a common expression across Turkish and English spoken in Göreme, used and understood to mean chasing foreign women. It is a term used by the local men in reference to their chasing tourists, and is also used with more negative connotations attached by tourists themselves in reference to their being chased either sexually or for their custom in restaurants or shops. In Göreme, though, the men working in tourism are by now well aware of the negative connotations attached in tourist discourse to the term 'hassle' and, as mentioned earlier, they pride themselves on their not hassling tourists to the same extent as men in other Turkish tourist destinations.

Some women's rejections of the attention they receive may stem from a more general desire to interact with the local people they meet on their trip in a way that somehow includes 'real' selves rather than mere stereotypes. One Australian woman, for example, told me in conversation:

> I'm not saying I'm cleverer than other women, but I can just see straight through the crap, I just don't trust them. It's all this 'I love you, you're beautiful' and so on, but I haven't fallen for it. They're always after something else, and I don't think that one of them isn't. I don't trust them. I could have gone for them, but I didn't want to get involved. And they're too intense – all this 'I'll kill myself' stuff.

Many women doubt that the attention a man shows towards them in the tourism domain is based on attraction and choice of them in particular, and reject the attentive advances they receive; this choice itself perhaps being experienced as a way of redressing the imbalance of power they sense in interactions with these seemingly overbearing men. The men's behaviour, on the other hand, is a response to the 'eye-washing' of the beautiful and free tourist women, and it is also a direct manifestation of certain aspects of the traditional gender relations in the village.

Traditional gender relations

The traditional gender relations in Göreme, and most other Central Anatolian villages, are such that men and women do not meet except with close kin or in marriage.[5] The women's domain is within the realm of the household, and to socialise in public, or *gezmek*, 'even with her husband', is possible only at particular formal occasions such as weddings or engagement parties. The behaviour of tourist women is thus deeply inappropriate to local ideas about gender identity and behaviour. Being a tourist is the ultimate in being out and about (*gezmek*), and so even before considering the behaviour of tourist women when they are actually in the village, the fact that many of them are travelling independently of their menfolk back home is a complete anomaly in the local view. The villagers have had to stretch the boundaries of their gender repertoires a long way to grasp the concept of touring women, and have succeeded in doing so to varying degrees and depending on the level of contact and experience they have had with tourists.

Villagers are generally able to separate themselves from the tourists on moral grounds. As I showed in Chapter 4, women's identity is primarily based on Islam. Knowing that the tourists are generally not Muslim enables them to position tourists clearly as 'other', and thus allows them to accept the tourists' uncovered hair or their short sleeves and trousers. To villagers, tourists are *giaours* (infidels) and whatever tourists do, whatever they wear, villagers know, or believe, that it is all right for them to do so in their own country. This ability to separate the *giaours* from themselves has enabled them largely to 'get used to' tourists' infidel behaviour:

> The people have got used to it. Everyone has really got used to it, they don't get uncomfortable anymore. But they say, for example, the very old ones, they say, 'Look, how they are coming, they are very young but they can come here. Our girls are by our sides all of the time, but they can come here. How do their families give them permission [*izin*]?' Some people talk like that. But for us it is not a problem, we've got used to it, to the tourists. And now my mother says sometimes, you know, when I wear jeans sometimes, she says: 'You *gezmek* like a *giaour*.'
>
> (Göremeli girl [*trans.*])

Only tourist women *gezmek*, and that indeed is precisely why the men are drawn to them. This was explained by a Göreme man who owns a tour agency in his telling me:

> In our eyes, in our heads, the women would always help the husband, everywhere – clean, cook – this is what we think about the woman. OK, you could take her out, but she doesn't want to go out, she doesn't like to go out. She is shy, because she hasn't eaten in a restaurant maybe all her life…Sometimes the man needs to do this because we are seeing it from

Europeans. They are very happy, having dinner together, going to a bar, drinking. They look very happy. I think we learn from these guys [the tourists]. Also we want to do things like that, what they're doing. So you go and ask your wife, she doesn't want to come, so you have to look for a girl. That's the reason to hassle girls.

The men are consequently learning new ways of relating to women, though of course they are well aware that their fun in the *pansiyons* and bars is always played out against the context of the village. The young men in particular are accused by their elders of turning their backs on their religion and tradition. One man explained:

> Formerly there was no tourism in our life. People were going to the gardens to work and adults were going to the mosque and children were going with them after school. Now what's going on?...If boys are with European girls and getting drunk in pubs, it is impossible for them to read the Koran and practise Islam.
>
> [*trans.*]

There is clearly a generation difference in the ways that local men are responding to and behaving in the tourism processes. Many of the middle-aged men, such as Abbas, who had their fun with tourist women ten or fifteen years ago, are today considered to have returned to a way of life more appropriate to village tradition. The younger men, on the other hand, who have only known tourism and who are growing up with the bars and plentiful 'available' tourist women, are increasingly drawn by the pulls that tourists and tourism present to them. They are drawn away from religion and also away from the codes regarding gender relations that are traditional in Turkish village society.

Marriage (*evlilik*) in Göreme is an arrangement made strictly between the families of the boy and girl, though the children themselves are increasingly being given a say in who they would like to marry, as they are in more urban regions of Turkey. In order that the girl's shame and the honour of her family be kept safely guarded until the day of her marriage, there is no 'courting' between unmarried boys and girls except perhaps for occasional chaperoned meetings between engaged partners. Ideally, marriage candidates are selected by parents and ultimately decided on by the patriarch of each family, the selection criteria being predominantly based around issues such as hard work and good temperament for a prospective bride, and family wealth and honour associated with the boy.

When asked whether love ever featured in choice of marriage partner, a typical reply from villagers was: 'No, if they're lucky, love will come later.' Younger villagers' answers were more mixed, however, suggesting an emerging ideal of romantic love in the dreams of adolescents. As with other societies where arranged marriage is the institutionalised norm, love, while not considered to be entirely separate from marriage, is not considered to be a primary reason for the marriage union.[6] Nevertheless, the concept of romantic love has long held a central place in Middle Eastern poetry (Magnarella 1974) and Turkish music is filled with the desperation of *kara-sevda* (doomed love or, literally, black love). Moreover, romantic love is becoming increasingly 'visible' for Göreme villagers, not only through the behaviour of tourists but also through exposure to Western films, television and travel/migration. When I visited the homes of Göreme women in the afternoons, they would often be sitting enthralled by a love entanglement being played out in a Turkish soap opera on TV. If I asked how the women felt as they watched life situations that were so different from their own, they shrugged and said that 'for others it is like that, but in Göreme it is like this'. Hence, while the beginnings of an ideal of romantic love seem not too distant, the parameters of emotion in traditional marriage rules remain firmly in place.

Furthermore, because of the strict codes of shame and honour, it is women who are kept most firmly within the parameters of traditional gender roles and relations. Men, on the other hand, particularly with their ready excuse of working in tourism, are relatively free from traditional village gender-codes while they are in the tourism realm. The men are not only drawn increasingly towards what is on offer to them in the tourism realm, but they are also expanding their repertoire of possibilities regarding gender relations. Through entertaining and socialising with tourists, the men are learning new patterns of courtship: they are going out with and socialising with women in a way that is not possible within traditional gender relations and in a way they had not done before. A young *pansiyon* worker who has a tourist girlfriend told me:

> Before we didn't have any chance. We couldn't go out with Turkish girls, we couldn't go to bars, we couldn't have fun, we couldn't meet each other, we couldn't know each other...Turkish girls are slowly going out – in Istanbul and Ankara they are, but not here, not in Göreme. But here it is also good, really. It is good to share everything with the tourists.

The men are thus learning *how* to go about courtship and having a girlfriend, so that increasingly what started off for both the man and the tourist as a part of the play and fun in the liminal tourism realm turns into something longer term.

Long-term relationships

The presence in Göreme of long-term tourist girlfriends has steadily increased in recent years. They stay for different amounts of time depending on the success of their relationships, and some women come back repeatedly from year to year after spending winters in places such as London where they can earn money to keep them throughout the following summer. A few tourist women have become permanent residents in the village, either marrying a villager or having long-term plans in that direction. The women are of various nationalities: many of them from Australia and New Zealand, others are from northern Europe, North America, South Africa and Japan. All of them work in tourism businesses: some investing in and running *pansiyons* with their partner, others earning their keep by serving in bars or sitting outside travel agencies or restaurants in order to 'catch' customers. Only very occasionally does a woman, without having a local boyfriend, stay and work in the village simply because she enjoys being there. Such women are usually 'hassled' so much that they either give up and start a relationship or they leave.

Following are some women's accounts of how they ended up longer term in Göreme:

> I stayed another week and then I had to catch a flight back to Sydney for my best friend's wedding, and Mustafa asked me to stay and I'm like 'I can't', and he said 'Well if you go, you're not going to come back cause I leave for the army in four months.'...And I went 'Ah, I'll just miss the wedding.' So I rang Rebecca...and I said to her 'Look, I've met this guy and I really think there's something huge between us, that I'm falling in love with him already.' But we'd only been here a week, but I said 'There's something big between us, he's asked me to stay, and I know you're going to be really disappointed but this is something I have to do, and I want to do.' And for the first time in my life I did something for me.

> We had all intentions of going to the Middle East until we got on the bus that morning. We just thought 'What are we doing on this bus?' And after two-and-a-half hours, we're were like, 'No, come on, let's go back.' The only thing that was stopping us from doing it was losing face with people we'd told we were going to do the Middle East. That was the only thing – which is just so dumb! Because travelling is all about meeting people, and that's what we'd done – we'd found people that we loved – and came back! And a lot of the people we'd met here seemed a lot closer than a lot of our friends in London, even my friends at home. Some friends you've known for life don't feel like friends like this.

Most women's accounts of why they stayed in or returned to Göreme combine an expression of romantic commitment to a particular man with an attraction to a lifestyle they perceive to be possible in Göreme as a place. As was discussed in Chapter 3, many of the Australian and New Zealander back-packers in particular are undertaking long trips in Europe, many for around two years in duration, before they embark on their life career. Unlike most of the northern European travellers, therefore, they have no fixed strings pulling them home after their holiday in Turkey. These longer-term travellers express a desire to escape from the drudgery and 'normal' expectations of a career back home. They are thus more open to the notion of diverting the path of their lives by exploring unusual and exotic possibilities. One Australian woman explained:

The concepts of what we've grown up with as a normal life – you work, you save, you buy your house, you buy your car, you get married, you have kids, those fundamental things that you're brought up with doing, getting your pension fund and everything like that. They are not like that here, they are not established, they don't know, that's why it is sort of not reality, because it's just different from everything you'd be doing at home.

An alternative life in Göreme is attractive not only because it promotes a sense of freedom, but also because it is a chance to purposefully reject norms and expectations present in the women's home lives. As one of the women quoted above said, 'for the first time in my life I did something for me.' They frequently express a pride in their having ignored or gone against parents' wishes and friends' warnings not to 'get involved with a guy from the Middle East'. They are actively rejecting over-protective and restrictive relationships in their home environment through their escape to and survival in a somewhat forbidden and exotic world. Thus these women's decisions to stay in Göreme clearly combine a romantic ideal about the life and love they might have there, together with a strategic *choice* concerning their own lives.

Local men too see their long-term involvement with a foreign girlfriend as something of an escape from the ties and restrictions surrounding traditional gender relations. One young man explained this in the following way:

I have a girlfriend, a foreign girlfriend, and we suit each other very well. So I don't think that I could find the same characteristics of her in a Turkish girl. The foreign girls think more freely than the Turkish girls, it is easy to communicate. And we don't care about the culture, tradition, religion. We don't care about any of them. But if I want a Turkish girl, it would start with her parents, her parents would have been involved in our relation, it's our tradition. Of course there are many reasons to be attracted to my girl-friend, but I knew also that nobody will be involved in our relation, neither her family, nor my parents. They can say something, they can try to be involved in our relation, but she told me and I told her that we don't care, we didn't care about anybody else.

The language in which these relationships are discussed is filled with notions of freedom, choice and the defiance of restrictive structures in place in both the women's and the men's home societies. By entering into a long-term romance with the foreign 'other', both the men and the women are at once embarking on something new and something perceived to be emancipating. As with the 'fun' part in the earlier days of a woman's stay, longer-term relationships continue to be played out largely in the tourism realm of the village, in the *pansiyons*, restaurants, shops and bars. The women frequently congregate together, providing familiar and easy company for each other, and discussing the latest 'drama', such as a fight that occurred the previous evening in the disco or a clampdown by the *jandarma* on their illegal work.

The young men, their boyfriends, have come to call them 'the local girls', indicating that they are no longer tourists or guests, and also suggesting that they should begin to adhere to village gender-codes. It is in this regard that, as the relationships progress, problems and conflicts emerge between the couples, as well as between men and in their relationships with each other. Tensions are clearly created between the 'traditional' and the 'new' regarding gender relations and ideals within the village context, and throughout the summer in particular there are often fights in the bars and discos in Göreme. Fights are sometimes against tourist men who are seen as interfering with a local man's chances with a tourist girl and sometimes between local men when one man sees his 'possession' of a particular girl being challenged by another man. Once belonging to a particular man, a woman's sexuality is potentially dangerous, as she has the power to provoke trouble between men. This was described by one man as follows:

> All the women are coming over – it's changing everything. Women came, they stayed here, and there are many problems going on in town, fighting and killing because of the tourists. It wasn't like this before. Tourist women, they sleep with another guy, and they sleep with another guy tomorrow and then they all have to face each other tomorrow. And it's starting to make problems for the guys – especially the young guys. They are active, they are young. They want to go out and they want to meet a woman, and it's getting worse and worse every year. It's no good. But you can't tell the women 'Stop doing this', and you can't tell the men 'Stop doing this'.

Unlike local women, tourist women have the power to choose, to reject and to play among local men. Unaware of the codes in the scheme of local gender relations, tourist women can behave in ways that provoke often violent disputes among the men, as well as misunderstandings between the woman herself and her boyfriend. An example of this came from a woman from South Africa. I met her one day in a *pansiyon*, and she told me of her experiences with a Göreme man she had been seeing for about ten days. She was very tense and told me almost immediately that she was 'having problems with a guy'. She said that the problems had started when she had gone for what she considered to be a harmless walk with a tourist guy she had met in her *pansiyon*. This had angered her Turkish boyfriend, though she did not understand why. She went on:

> He was very charming for the first few days, but now he's turned *very* possessive. He's treating me like a possession. He won't let me go out, he tells me to sit and shut up, what to wear. He told me to change before I went out to the bar one night – told me I couldn't go out like that here. He won't let me smoke, he told me not to talk while we're eating. He won't let me go out alone – even down the street. It's archaic, he's really a

peasant! He's got all his spies out. They all know that I'm with him, he's told everyone, and he says 'This is my town', with the idea that I have to do what he says or else.

Misunderstandings of this kind frequently occur between tourist women and Göreme men. As soon as women are considered to be attached to a man, their interaction with other people in the village becomes limited and they find themselves subjected to rules and conditions that they do not understand. These are the rules and conditions, or a confused version of them at least, that exist within the context of local gender relations, and they not only restrict the sense of play and freedom that the tourists initially expected from their stay in Göreme, but can lead to disputes between partners. The South African woman quoted above had, from the 'local' perspective, behaved in a way that would be potentially damaging to her boyfriend's pride and honour. He had therefore acted towards saving his pride and honour, which involved playing a heavy hand and subjecting her to his – and the village's – rules. He acted to remove any sense of power her sexuality might have had, leaving her feeling sour and with no choice but to leave the village. Similar situations occur time and time again throughout the summers in Göreme, some happening a few days into the relationship and some after a few months. As a further illustration, a Canadian woman talked of her experiences like this:

> It was fun when I came. I had fun with him. He never made a pass at me and we used to go for walks, and then we'd go out for dinner, and then we'd dance all night. But then it became work. I think at first Turkish guys are attracted to the free spirit of the foreign women, but then they start to impose rules, like don't wear short T-shirts, so it cramps you…and now it's lost its charm, it's worn thin. They become more controlling of you, and then they go out and do things with other people that they used to do with you, and leave you to do all the work.

Such situations, or their outcomes at least, lie behind the accounts of touristic sexual liaisons between local men and tourist women elsewhere that have portrayed the men as strategically abusing their tourist 'victims'. Zinovieff (1991), for example, places a strong emphasis on the way Greek men cast women out after the conquest, since they are using their sexual conquests with tourist women largely as armour in their competitive relationships among their peers. In Göreme too, games and competitions occur among the men regarding their sexual conquests in the tourism realm. I heard groups of men judging newly arrived girls on whether they were likely to be 'easy' or 'difficult' (to get into bed) in order to then make bets with higher kudos for scoring a 'difficult' one. I also heard men teaming up to go out to the bars to get 'chicks' for the night. Some men and boys in the village have achieved higher status among the others because of their skill and the number of their accomplishments in this sphere.

However, it was pointed out by some men that competitions and showing off has lessened during the past few years, because tourist women have gradually become more plentiful and so it has become more common or 'normal' to have sexual relations with them. With the exception of the younger men, for whom these activities are new and exploratory and so still play a role in achieving status among peers, if men tell each other what they did with a tourist girl last night, the reply is 'So what?' The relationships have thus become valued in themselves rather than being wholly part of a male system of prestige. When the relationships break down then, such as the situation described above of the South African woman, it is not necessarily because they were always intended as short acts of conquest for the men concerned. Rather, it is because of tensions and conflicts emerging from clashes in the codes and understandings concerning gender roles and relations between the two partners.

It is perhaps because of their concentration on male narratives 'after the event', therefore, that the portrayals of such relationships from Bowman (1989) and Zinovieff (1991) repeatedly place a template of rational strategy over male behaviour and contrast that to the 'weakness and femininity' associated with emotion (see Seidler 1987). Bowman (1989) tells of the way groups of Palestinian men in Jerusalem – who are 'feminized', thus weakened, by their economic and political position – are able to regain a 'masculine' position through sexually dominating the women of the 'dominators'. It is certainly likely that as a single woman researcher in Göreme I was unable to obtain quite the same male narratives of sexual conquests as Bowman could in Jerusalem. Moreover, there is no doubt that, as I described in the previous chapter, the negotiation of power and 'rights' between tourists and their hosts is always in process. The men's assertions of control over their tourist girlfriends, therefore, may be similar to the villagers' broader assertions of control over the tourists in *their* village generally, as achieved through the pronouncing of tourists as 'guests'. Furthermore, parallels may be drawn between Bowman's Palestinians and the men working in Göreme's tourism businesses who come from other parts of Turkey, especially Kurdish men from the south-east. Those men undoubtedly experience disempowerment regarding the Turkish political arena and, as outsiders, they are also in a weak position relative to Göremeli villagers. It is interesting to note that these men were generally the group in Göreme who were most strongly accused of 'ripping off' foreign women in village male discourse.

Nevertheless, it is doubtful that relations with tourist women function in the same way for the Göreme men. Being on home ground in their village, their pride and power in gender as well as economic relations is not so evidently in the balance as is the situation for 'outsider' men. The Göremeli men thus seem to be exploring the new experiences of charming, courting and socialising with women. Concurrently, however, as the relationships become longer term, the tensions and conflicts experienced in the juggling of relations and gender codes between the tourism and the 'back' realms of the village grow more intense.

Caught in the middle

Since these tourism relationships are a 'new' development, there arises the uncertainty and the lack of parameters and codes of practice that accompany any form of social change. As a couple's relationship becomes longer term, for example, the women have certain ideals and expectations regarding how much time they should spend together and how close they should live their lives. The men too may develop ideals of living and sharing time with their girlfriends, but they are less able to express these desires because of the pressure of village behavioural codes on them. Thus the men find themselves caught up in the tensions that exist between the tourism and traditional realms:

> In Göreme everybody knows each other, it's too small, and also our fami-
> lies look at us. Normally in the house we never touch, we never kiss, they
> never sit close. They never touch in the home with family. For example, if I
> was married, and my wife was sitting here, we would never touch because
> they see it. It doesn't look good. That's shameful for us, shame...In
> Göreme, we can't walk together in the centre, or near the café, because
> everyone will see us. The old people, they can't do anything to us, but they
> tell people. It's shameful. They tell that he is having a tourist girl, they are
> gossiping. And they are saying to my father and mother 'How will you find
> a Turkish girl for him?', because I am all the time going with tourists.

Gossip and shame act as strong social controls in Göreme society, and the men experience the intensities of this form of control when they contravene village tradition by 'all the time going with tourists'. The men feel somewhat torn between the values and expectations of the tourism realm and the 'back' of the village.

Further tensions arise between the tourist women and the traditional realm, and also between the local women and the tourism realm. The social controls of teasing and gossip, as well as the occasional firmer reprimand, press the tourist women into having some awareness of the many rules and expectations of them in the village. With their activities occurring mainly in the tourism realm, tourist girlfriends usually maintain something of an uncomfortable awareness of the 'back' areas of the village and the women who are related to their partners. This was evident from the tourists' frequent questioning to me, in my position as a link between the two realms, concerning what the local women thought of their presence in the village.

A tourist woman's respect among the people of Göreme is lacking from the outset because of her being 'open' (*açık*) and associated with infidelity. Many of the women's reticence to learn Turkish is another barrier in their communica-tion with villagers, and if they wish to remain living in the village in the long-term, they must also work hard at gaining respect through learning to behave appropriately within the village. Some of the women do try, however uncomfortable they feel, to spend some time with their partner's family. Others are shamed into keeping away, feeling much more comfortable in the tourist sphere of Göreme. When I told one Australian girl that I had just been speaking with her boyfriend's mother, she assumed that the mother must have said bad things about her. I asked her why she assumed this. 'Because I never go up there – they're too scary!', she answered.

I had met this mother while I was with a group of women making bread in their neighbourhood. I had asked them what they thought about all the tourist girls coming and staying in Göreme. 'We don't like it!', the mother exclaimed.

> Would you like it if your children went with 'others'? We can't get on. We have a lot of work, bread, grapes – we are always working, and our men are going around with tourist girls!

She expressed concern that she and her son's tourist girlfriend could not understand each other, and so would not get along in the future in the same way that she would with a Turkish daughter-in-law. She was concerned in the same way for her son and the danger for him in the new and unknown quantity of his long-term involvement with a tourist girl. Unlike Turkish daughters-in-law, tourist girlfriends can always leave, and she talked of this and other young men's hurt in the past when their girlfriends had left them and not returned. The concerns of these women emphasised the ideas held by villagers that foreigners are dangerous and threatening to their relatively closed village order.[7]

Yet there have been approximately twenty marriages to foreign/tourist women over the past fifteen years, and the numbers continue to increase by one or two every year. Marriage to a foreigner is more or less condoned within the village. Brides are, in any case, very often incomers from other places, be they other villages, provinces or countries (through migration). However, tourist women are recognised as being unable or unwilling to come into their husband's family as a Turkish daughter-in-law would, and as a partaker of women's work within the household. The structures of households and gender relations are therefore changing significantly, though it is important to note that changes in family and household structures were already occurring through outward migration from the village.

Complaints, heard particularly from elderly people in the village, concerning the young men's relationships with tourists tend to refer more to a general absence of marriages, and thus brides (*gelin*), due to the men's new-found play and courtship with tourist women. The concern is that men are marrying at an increasingly late age, and households are consequently left with no *gelin* for an elongated duration. Another concern is that men who work in tourism and experience the 'fun' of foreign girls will carry on with this play even if they do comply with their parents' wishes to marry. As the young man quoted above said about his situation, 'They are saying to my father and mother "How will you find a Turkish girl for him?", because I am all the time going with tourists.'

Through networks of gossip, women in Göreme are well aware of what takes place in the tourism realm. However, while men's relationships with foreign women might be socially problematic, they may be more accepted by women and the village as a whole on economic grounds. Women accept their husbands' staying out late and going to discos with tourists largely in the understanding that they are working. It has been learned through the past years of running tourism businesses in the village that tourism is about entertaining people, and that might necessarily involve taking tourists out to dinner or for a dance at the bar. The stretching of socio-cultural boundaries is thus justified, to an extent, for economic reasons. Some foreign women told me of being invited to their boyfriend's house and meeting his wife. A few foreign women have even moved into the man's home, or *pansiyon*, to live with him and his wife. The man would usually tell his tourist girlfriend that he had been forced into the marriage by his family when he was very young and that he had never loved his wife and no longer had relations with her. That seems to satisfy the foreign women who, because of their own ideals of love-marriage, consider his loveless marriage to be void. The greatest problems come, of course, when men who are married to a foreign woman continue to behave this way and carry on going out with other tourists.

Men married to Turkish women, for the most part, receive less 'trouble' from their wife when they play with tourist women. Although conversations among village women about their husbands' infidelity are expressive of some contempt, their tone generally remains light-hearted and jovial. Since marriage in village tradition does not primarily include an ideal of love, jealousy in the sense that many of the tourist women imagine to be an issue is not prevalent. This was confirmed by a village girl who, in conversation about marriage and jealousy, told me:

> Everyone marries, but they don't know love, they can't find love, they cannot love them, because they are not a good person, but they are obliged to live together. For that reason they do not get jealous. Whatever is done, let it be done. It is not important at all in that case.
>
> [*trans.*]

Hence, the men's 'playing' with tourists for short-term flings is generally tolerated.

However, the gossip that ensues about such relationships can be more hurtful to village women than the actual behaviour of their husbands. When I asked one villager, for example, what the Göreme women thought about the tourist women having relationships in the village, she told me:

> They all ask 'Will they marry, will she take your husband?' – They said it about me a lot. My husband went out with foreign women, especially one for a long time, and everyone said 'He will marry her.' But I knew he wouldn't, because we're happy like this. And we became friends, me and the girl, we slept in the house together, we ate together, I liked her a lot. But all the gossip – everyone saying she'll take her husband from her – that she wouldn't go. I know he did everything, I see everything and I know everything, of course I do.
>
> [*trans.*]

These words also indicate the village women's fears concerning the possibility that their husband might actually leave them to be with a foreigner more permanently. It is largely accepted that the men will have their play with tourist women, but it becomes a different story altogether when a woman's husband develops a longer-term relationship that might continue into the future. Because women are completely dependent financially on their husbands, it is devastating for her and her children if the husband leaves to be with another woman. On this, one village girl said:

> They can't say anything, because when they get divorced their family doesn't want to take them back. It is difficult, very difficult, because their family will ask what happened, and they will say my husband was unfaithful to me. 'That's natural' their family will say. 'It is normal' they will say. So they cannot come back, women cannot divorce. It is very difficult.
>
> [*trans.*]

The foreign women are not necessarily blamed for these occurrences. As one girl, whose father had left to live with a German woman some years ago, said:

> We don't like them [foreign women], because they are breaking up families, but many of the men in Göreme don't tell them that they are married. Then by the time she finds out, she either doesn't believe it or she doesn't care.
>
> [*trans.*]

It is highly dishonourable for a man to leave his family in this way. Hence the attempts by some men to bring together their wife and their girlfriend under one roof: he can then have the best of both worlds. Likewise, the (economic) devastation caused to a woman if her husband leaves her explains her acceptance of a foreign girlfriend into the home. What takes place within a marriage tends to be less concerned with emotion and more with the economics of the situation.

Developing business through romance

Foreign women are sometimes welcomed into families largely as providers of wealth, either directly through their investments and work in tourism businesses, or as offering the men they marry the opportunity of going to the foreign woman's home country to work. As it was noted earlier, local entrepreneurs have come to believe that they will sell more rooms, meals, tours and so on if they have a tourist woman working for them to 'catch' potential customers. Similarly, many of the long-term tourist women are asked by either their boyfriends or by other men to invest money and enter into partnership in their tourism business. In contrast to local women, then, whose value is in their domestic and garden work and procreation capabilities, foreign women are forging new gender roles in the village where their presence, work and investments are seen as increasing the economic opportunities of tourism business.

Many of Göreme's tourism businesses were built on tourists' money or work. This process began with the first marriage between a villager and a tourist woman fifteen years ago. The ideas and styles of these often successful businesses are used as prototypes by other entrepreneurs. Examples of style elements introduced to *pansiyons* by 'tourist managers' include: dormitory rooms that are cheaper to stay in than double rooms; communal areas with floor cushions where tourists can 'hang around' and meet with each other; laundry services; book exchange systems, and so on. Tourists have also opened cafés that sell cappuccino coffee, chocolate brownies and vegemite sandwiches. Turks who work in these businesses learn how to make these 'tourist foods' and may later open their own business selling similar items. Longer-term tourists are therefore frequently the innovators in Göreme's tourism business.

It was noted earlier that the tourist women's narratives concerning their staying and investing their lives in 'romantic' Göreme demonstrate how they made an active choice to do something for themselves. They are attempting to improve their own lives by escaping the social and financial pressures that they perceive to be present in their home environment. The longer they stay in Göreme and the more involved they become with a Göreme man, however, the more they inevitably become involved in the social and financial pressures associated with their partner's home life. One woman who runs her partner's *pansiyon* said, for example:

> His whole family is taking money out of my pocket. I'm getting used for work. I mean I'm not doing anything that I wouldn't normally do anyway, but when it comes to being fair, I'm putting all the hard labour in, and then most of it goes to his family.

Another woman, who was working in a bar unconnected with her partner's business, said:

> I know it's their duty to look after their family and give them money and all that, but I find it very hard 'cause if he's got any money it goes to his family. He's never got any money, 'cause he's paying for them, and I find that very hard to cope with. So that's what makes me think that I can't come back here and work and just live to support his family, 'cause that's not the way it works for me. You work for yourself, and you work for your kids' education, or to make your business better or whatever, but with his family, it's never going to work like that.

In other cases, tourist women have invested quite large sums of money in the businesses of their partners. A problem for them is that they cannot obtain any legal status regarding their investment and, if their relationship later breaks up, they have little power to take back their financial investments. What had started as a woman deciding and choosing to invest in bettering her own lifestyle often turns later into a situation in which she feels trapped and largely disempowered in relation to both her partner and the wider Göreme context that she is in. One Australian woman, who had been in the village for two years, told me that:

> In the short-term relationships, the tourists are in control, because they're here for a short time and then they're off. But in the long term, the men are in control because somewhere along the line it enters into some sort of business relationship. And whether it's cultural or financial or whatever, the men tend to control. Even women who I'd thought of as being fairly strong seem to be dominated by the men in that sense.

This woman was reflecting on her own situation, in which she feared having lost tens of thousands of dollars to a villager. From the villagers' point of view, the fact that many women have entered into these joint business ventures has led to a belief that, when capital is lacking, the simple answer to building up a tourism business is to: 'Just meet a tourist girl and get her to invest in a business together, telling her that you can make good money together.'

The village women too, while being marginalised through the men's romantic liaisons with tourists, are tolerant for the sake of potential economic gain. It has been frequently noted in discussions on gender and development that, as the development of capitalism and entrepreneurship occurs, traditional gender roles limit women's access and rights to any property and business for themselves, thus serving to reinforce women's dependency on men (Scott 1997; Sinclair 1997; Starr 1984). This has certainly been the case in Göreme with the growth of small tourism enterprises. It follows that a way for Göreme women to include themselves and to have more control over the economics of the household might be to gain a daughter-in-law (over whom the mother-in-law has most direct and everyday control) who does have financial resources. Thus a local woman's access to financial resources may be gained through the 'touristic liaisons' of her son or, in some cases, even her husband. So, as mentioned above, while these touristic relationships might be socially problematic for local women and villagers generally, they may be tolerated or even sought for economic reasons. This point was demonstrated in encounters I had with some of the poorer families of the village: mothers would suggest that I (as a representative of tourist/Western wealth) married their sons, and on one (rather unpleasant) occasion it became clear that a wife was trying to place me together with her husband. It is always expected that any honourable man will provide for his family, and so it is not only for the men concerned that foreign women are seen as being rich and a possible route to economic salvation.

In addition, as with most other case-study accounts of sexual/romantic liaisons between tourist women and local men, marriage to a foreign woman might be seen by the male partner, and possibly his family also, as a means of escape to the woman's home country and a prosperous future.[8] Unlike the situation in the 1960s and 1970s, it is fairly difficult today for villagers to obtain a visa and legitimately migrate to northern Europe or another Western country. Marriage to a foreigner makes this a little easier, and one day I heard a group of teenage boys wandering through the streets singing 'No woman, no visa' to the tune of Bob Marley's 'No woman, no cry'.

These, of course, are the key scenarios that give rise to popular as well as anthropological portrayals of the men in these relationships as the strategic 'players' and the women, stricken by romance, as the hapless victims of the men's exploitative tactics. While these men do dream of escaping to a richer land, and soon learn from others around them that to court and then marry a foreign woman is the easiest way to achieve this, it should be remembered that alongside this the foreign women are also acting out of choice and adopting a strategy towards a happy future.

In more recent years, however, with a number of village men having been to northern Europe and Australia to work with their foreign wives, the view of prosperity in a foreign land does not appear as rose-coloured as it used to. Many young men have returned with stories of finding it hard to gain employment there, and also of being treated badly among a hard and cold people. Along with this disillusionment often comes the breakdown of the marriage, so the men have returned alone and disappointed. With tourism continuing to develop in their home village, then, it is gradually becoming considered a better prospect to stay and, if need be, obtain a foreign partner's help in the starting and running of a tourism business in the village. Since the tourists are often in an economically stronger position than their partners, they want to invest in their life in the village together with their partner:

> If you're going to make a life together, that's normal isn't it? I mean, you get married, and most people get married for life, so you think 'this is my life', so you're going to put money in.

Today, then, after stories of unhappiness and lack of work in those foreign lands from where the tourists come, many couples are choosing to stay in Göreme to build businesses together in tourism. Along with the innovative ideas these foreign residents have regarding tourism business, and the changes to the 'landscape' of Göreme's tourism that result from those ideas, the women's work in this realm is demonstrating and creating a new type of gender role in the village. Instead of entering their husband's home and working within the household in the traditional way alongside the mother-in-law, these new wives are working together with their husbands in tourism business. The husband's family gains economically from their new type of *gelin* (bride), not from her work within the household but from the financial gains she brings in from the tourism realm. Her work in the tourism realm is condoned because it is understood that that is what she knows best. After all, she is a tourist herself.

Moreover, the tourist women themselves, because they are working in tourism, find a reasonably comfortable place halfway between the tourism realm and the traditional realm. While they are expected to spend some time with their female relatives and to attend family occasions such as weddings and religious holidays, whenever they wish to escape back to their more familiar cultural environment they have the ready-made excuse that there are tourist customers to attend to. The presence of foreign women and the romantic relations they develop together with local men are thus forging new ideas about gender in the village: from their beginnings as romantic liaisons, where young local men are learning about the possibility of courtship and romance, through to marriage and the development of new business in the village, along with a new 'type' of and role for the *gelin* (bride).

Developing romance: changing village life

In this chapter I have discussed the ways in which gender identities and ideologies are re-negotiated and moulded at different points of the intimate relationships between Göreme men and tourist women. Rather than merely studying partners' retrospective narratives after relationships have taken place, which tend to emphasise misunderstanding and abusive power relationships (usually in favour of the men), I have viewed the relationships as processes throughout which the expectations and power of each partner are constantly negotiated. Looking at how the relationships begin in the fun and playful context of the tourism realm, and following them through to longer-lasting relationships and marriage, allows us to see the choices and strategies of both the men and women involved, together with the part that the anti-strategy of romance and emotion plays throughout. By re-negotiating gender conditions set in their home context, men in Göreme are experiencing and developing a new taste for romance, as well as exploring ways of using that romance to develop tourism business in the village. Similarly, foreign women are *choosing* to stay and negotiate new roles for themselves in the village, new roles that themselves challenge and re-work the values and ideologies regarding gender identities throughout the wider context of Göreme society.

Concurrently, problems and misunderstandings do occur between 'romance' partners because of clashing concepts of gender identities and expectations. As the foreign women's status changes from that of 'tourist' to 'insider', so it becomes necessary and expected that her behaviour corresponds with the traditional gender relations of the village. Foreign women therefore consider themselves to be more empowered in the earlier stages of their relationships. As the relationships move towards the possibility of marriage and lifelong commitment, the identity and status of each partner in relation to the other, as well as their status in relation to the context of gender ideologies in the village, can become problematic.

Some young men are learning from these difficulties and deciding, after years of fun and play with tourists, to marry a Turkish girl. One young man who, after many tourist girlfriends, had finally married a girl from the village, told me:

Maybe we've lost some of our culture, our traditions, but still we have some. I've lived with tourists all this time, learned about them, their very different culture – even if I have some of your culture now, it's very different. You can't tell a Western woman 'No, you must stay at home, you can't go alone to the disco.' Anyway, they don't listen, they just leave.

This man, however, married a village girl whom he chose carefully for her non-typicality regarding the usual conformity to gender identity in Göreme. He wished to marry a girl who understood village expectations but who would also be able to fit in with what he had learned regarding courtship and socialising through his relations in the tourism sphere. Many men in their twenties are now following a similar path, choosing to marry a Göreme girl but one who is more educated and 'open' because she had grown up in a Turkish city or in northern Europe after her family's migration. These men have realised that marriage to a foreigner is likely to eventually break down. However, they also wish to marry someone who might *gezmek* together with them, eating in restaurants, going on holiday and even travelling abroad.

The Göreme men's gender repertoires have thus expanded to include a blend of both local and tourist gender ideologies, a blend that is also serving to gradually re-work gender roles and identities of local women in the village. Through the romantic developments between tourist women and local men then, tourism is creating something very new in Göreme, not only through change in terms of economics and livelihood, but in the gender roles and relations that are at the very centre of villagers' lives. Such change is an inevitable manifestation of tourism in the village. In the next chapter, I will consider the implications of change in Göreme regarding the continuation of Göreme as a 'tourist site'.

Notes

1   A discussion of the use of the term 'romance tourism' versus 'sex tourism' to describe such relationships is provided by Herold, Garcia and DeMoya (2001). Previously, Pruitt and LaFont (1995) had chosen the term 'romance tourism' to describe the relationships that tourist women had on holiday in Jamaica, as opposed to the 'sex tourism' more widely discussed in relation to tourist men (Cincone 1988; Hall 1992; Lea 1988).

2   Kohn raises this issue with reference to inter-ethnic marriage in Nepal, pointing out that the anthropological literature on marriage shows a neo-functional leaning, leaving no room for the 'simple attraction' of the exotic other. Commenting on Bourdieu's account of the 'game' of marriage (Bourdieu 1990), Kohn argues that:

> the whole emphasis on strategy as the impetus for marriage does not leave room for the aesthetic spark, the romantic and wholly reckless anti-strategy of love, especially across culturally constructed 'boundaries'.
>
> (Kohn 1998: 69)

3   As Pruitt and LaFont (1995) and Meisch (1995) point out with reference to romantic relations in Jamaica and Ecuador respectively, close liaisons with a local man may be viewed by some tourist women as a key to her own access to local culture:

'What could be more back stage, and offer a more intimate experience of a culture, than being invited into someone's bedroom and bed?' (Meisch 1995: 452).

4   The Mayor of Göreme often used language of magic and bewitchment to explain the presence of so many tourist girlfriends and brides in the village. Kohn (1997), citing Schneider (1993), also discusses the bewitching forces that draw tourists into gradual residency on a Scottish island.

5   It is important to note that rural village life makes up approximately half of modern Turkish society, and that there are vast differences between marriage and gender practices and ideologies between the rural and the urban settings. Göreme is said by villagers to be extremely 'conservative' with respect to gender relations, even in comparison to Avanos town, which is only ten kilometres away. Today, young 'courting' couples are a common sight around university campuses, and in cafés and parks in Turkish urban society.

6   See, for example, Fruzzetti (1982) and Trawick (1990).

7   Delaney emphasises this point, referring to Turkish villagers' notions of the village as *kapali* (closed) and *temiz* (clean), just as a 'proper' woman should be (Delaney 1991: 207).

8   See Cohen (1971), Meisch (1995), Pruitt and LaFont (1995) and Zinovieff (1991).

8 The continuation of Göreme as a 'tourist site'

Politics of place and identity

Belgian tourist: Göreme isn't beautiful anymore. There are all the fairy chimneys and caves, which is nice, but everything is in English, and all the boards and names of the *pansiyons*...Things like the Flintstones Cave Bar, I don't like it – it's for more typical tourists...I suppose it's good to have for those tourists who like it, but also it's good to have more natural places...In ten years' time, there will be more signs, neon, it will lose everything that *we* come for.

Ministry of Culture Official, Preservations Office, Nevşehir: The local people are not cultured; they are villagers, they are uneducated. They only want to sell things; they don't know what tourists really want. Tourists don't want to buy carpets, they want to see history and culture. The Göreme people don't understand our preservation project because they are uncultured. We can't teach them how to protect the place – they make things ugly.

[*trans.*]

Göremeli villager/*pansiyon* owner: If I was mayor, I would restore the whole of Göreme, and look after all the old people here, that's more important. My friends come from other towns and they say 'Hey, Göreme looks nice, all the roads and flowers', and I think 'Oh, it's so stupid.' We don't need asphalt, we don't need flowers...All the rocks are falling down and the people have a lot of problems. If I was mayor, I would help those people first, and the second thing, tourism.

Underpinning all of the above quotations is the question of Göreme's future, though the meanings inherent in this question, and the meanings attached to tourism, differ for each of these interested viewpoints. Through tourism, places are imbued with multiple simultaneous meanings that might be both contradictory and contentious. For the tourist quoted above, the tension embedded in the Göreme landscape as it is commoditised for tourism is conveyed as the contradiction between the caves and the tourist signboards, between the natural and the neon. There is also obvious contention over the meanings of the Göreme landscape between the Ministry of Culture representative and the Göreme villager quoted above. Indeed, rather than being an 'empty meeting ground' (MacCannell 1992), Göreme has become a hotbed of communication between global processes and local concerns, and is thus a place full of vibrant negotiation concerning the future of the village and tourism there. This chapter is intended to draw the main themes and issues that have been discussed in the previous chapters together, in order to consider the continuation, or 'sustainability', of Göreme as a tourist site. I will first discuss the politics of preservation in Göreme, looking at the way that the discourse of preservation renders place and identity static and does not allow for change to take place. I will then go on to discuss the ways that the villagers and the tourists, through their interactions and play with identity, are able to negotiate and move beyond that static notion of place and identity.

The discourse of preservation

We saw in Chapter 2 that the aspects of Cappadocia marked off in tourism images and myths are its 'lunar' landscape, its Christian (Byzantine) history and the contemporary troglodyte way of life in villages such as Göreme. The growth of tourism has led to an aesthetic valuing of all of these features, and has hence served to promote their preservation. Much of the preservation and restoration work, such as retouching frescoes and filling cracks in the rock to prevent rainwater from further weakening the structure, is focused on the caved Byzantine churches in and around the Göreme Open-Air Museum site, and is funded jointly by the Turkish Ministry of Culture and UNESCO.

There is also interest, however, in preserving the contemporary cave-habitation in the villages of the area. This was seen in the extract from a Göreme National Park leaflet quoted in Chapter 2, and is clear from the original 'Park Plan' prepared by members of the United States National Park Service Planning Team on assignment to the Turkish government in the late 1960s. The plan described the 'vital living landscape with deep traditional ties to the spectacular "Chimneys of the Fairies" and cliffs that define the area',[1] and decreed that the villagers should be allowed to continue habitation in their cave-houses.

During the 1960s and 1970s, however, just prior to the development of tourist interest in the contemporary cave-life of the region, a government policy (AFET/Disaster Relief Directorate) was enacted to subsidise villagers' re-housing from cave-houses and into more 'modern' concrete housing. Deemed too dangerous for habitation because of erosion and threat of collapse, many cave-dwellings and 'fairy-chimney' houses were evacuated and their residents re-housed in government-built dwellings at the lower, flatter end of the village. Whole sections of Göreme village were declared 'disaster zones' and appropriated under the national Disaster Relief Directorate. At that time a general move towards more modern and prestigious housing was instigated, and that lower part of the village continues to be the main 'building zone'.

When tourism really got under way during the late 1980s, however, many of the re-housed villagers began to reclaim and restore their old homes for the purpose of making *pansiyon* businesses. Although all evacuated houses officially belong to the state treasury, this kind of activity has been tolerated because it has meant that such old properties are restored and maintained. This tolerance,

Plate 8.1 The National Park plan in action: municipality workers placing telephone and electric cables underground so that, in accordance with the Göreme National Park plan, wires and pylons do not obscure the 'natural' landscape.

along with the recent removal by the Department of Infrastructure of the 'disaster zones', is indicative of an increasingly pervasive interest in the preservation of the older part of the village.

The Romantic – and orientalist (Said 1978) – view of the contemporary cave-life is now manifested in the Göreme National Park and regional Preservations Committee (under the national Ministry of Culture),[2] both set up in the 1980s with the broad aim of preserving and restoring the historical and cultural heritage of the area. Although villagers have been allowed to remain living in their cave-homes, all rock structures within the National Park, which include many villagers' cave-houses, have been taken into government control and strict regulations have been put in place forbidding any unauthorised alteration to the existing rock structure, such as fairy chimneys or cave-homes, and also any unauthorised new building work. For anybody to carry out such work, plans must be drawn up and submitted to both the Municipality Office and the Preservations Office in Nevşehir. If alterations are carried out to any rock structure without obtaining the correct permission, the perpetrator may be subject to fines or imprisonment.

These regulations can be viewed as the regional filter of a more global preservation rhetoric that became institutionally formalised through efforts such as the UNESCO World Heritage Convention in the early 1970s, which decreed the need to preserve 'cultural landscapes of universal value' (Plachter and Rossler 1995: 15). Cultural landscapes of value are socially and politically constructed,[3] and as such frequently become points of contention between local, national and international levels of discourse. Further illustrations of global preservation rhetoric working at the local level in Göreme include the setting up of a 'Save Göreme Committee' by a Dutch man who has managed a *pansiyon* in the village for many years, and also a section in a recent version of the *Lonely Planet* guidebook entitled 'A Future for Göreme?' They are both aimed at promoting awareness among the visiting backpackers of the cause of Göreme's preservation as part of the overall pursuit of the sustainability of Göreme as a tourist site.[4] Such 'socio-environmental movements' (Mowforth and Munt 1998) are undoubtedly a manifestation of Western visitors' Romantic and orientalist tendencies, which advocate environmental and cultural preservation as ends in themselves. As Mowforth and Munt (1998) point out, these global socio-environmentalist movements and their associated tourisms (green, eco- and 'soft' tourism) are hegemonic in themselves, in that they promote these values as global *needs* and in turn are blatantly neglectful of local voices.

Similar values were frequently expressed by tourists I talked with in the village. An Australian woman told me, for example, that:

> [she] would hate the Göreme people to all be driving cars in twenty years. Donkeys and horses and carts are much nicer. It's nice for time to stand still in some places.

One middle-aged woman (though originally from urban Turkey, she was married to an Englishman and had been running offshore business from Guernsey for many years) who I got chatting to about my study in Göreme immediately exclaimed: 'Negative! Tourism is all negative! It just ruins people and places!' We were at a tourists' full-moon barbecue organised by a village tour-agency in one of Göreme's fairy-chimney-filled valleys; when she said this, the others standing around the fire – mostly younger backpackers – began to look slightly uncomfortable. They did even more so when I said: 'Well, in that case, everyone stop being tourists and go home right now.' There was a stony silence. This silence was indicative of the discomfort caused by their being labelled tourists, and also of the responsibility they had to admit regarding the 'ruin', as it was being described, that tourism and thus they themselves bring to places like Göreme.

The Save Göreme Committee and the article entitled 'A Future for Göreme?' are also riddled with paradoxes and contradictions. Besides that piece in the *Lonely Planet*, the writer also wrote an article of the same title for *In Focus* magazine, the publication of the London-based organisation 'Tourism Concern'. In this article she was scathing of local entrepreneurs who have 'eyes only to turning a quick buck and landing themselves a tourist girlfriend' (Yale 1996). Moreover, hearing of some of the politics surrounding tourism in Göreme, she was led to 'tremble for Göreme's future'. She also acknowledged in this article the contradictions between her 'personal feelings' and 'concern for Göreme's well-being' in the fact that she writes for the guidebook that is arguably one of the most powerful forces bringing tourists to Göreme today.

The advocate of the Save Göreme committee is engaged in similar contradictions, though for him profits from current tourism should be used to restore the old village, rather than the AFET and tourism organisations continuing to construct new buildings 'which are spoiling the natural beauty of the environment' (Leyssen and Idiz 1993). He argues that rather than being moved out into safer new houses, the villagers should be kept in their 'chimney' houses so that the houses and the culture contained therein will be retained. If the houses are neglected, they will go to ruin and 'this natural and architecturally historical unique monument – a natural open-air museum – will no longer be around for future generations to see' (ibid.).

We are reminded here of the 'trope of the vanishing primitive' illustrated in the anecdote concerning *pekmez* in Chapter 3. Indeed, both tourist discourse and official rhetoric expounding the need for cultural preservation in Göreme show a marked desire for the village to somehow remain static in order to suit their aesthetic ideals of Göreme 'village (cave) life'.

Villagers' experiences of preservation discourse

The aestheticisation of the 'old' village is gradually filtering through to villager discourse. The process whereby villagers have taken to restoring and making tourist accommodation from their previously abandoned cave-homes is indicative of an increasing awareness of the aesthetic and economic value of their cave-houses and cave-life. It is generally cheaper and easier to build and move into the new area at the lower end of the village because of the lengthy and difficult process of obtaining permission from the Preservations Committee to make any alteration to houses in the old sector of the village. However, the old sector is gradually being given a new lease of life, with more and more properties being bought and sold for residential as well as tourism business purposes. A trend is occurring whereby those villagers who have by now made some financial capital from tourism, and also foreigners or long-term tourists who seek a Romantic retreat in Göreme, are restoring and moving into the old cave-houses.[5] Indeed, a discourse of preservation is developing among villagers:

> We need to keep Göreme. We have to try not to destroy Göreme, we have a lot of future in Göreme. I mean, all of Turkey has a future in Göreme – I'm not just thinking about ourselves and profit...but the old houses and caves are what's really important. If you build new buildings then I think I'm going to do another business – it's really sad what is happening to this place now.
>
> (Göremeli *pansiyon* owner)

Similarly, Abbas told me:

> It was better before, ten or fifteen years ago, because it was more like real Göreme – the old life, donkeys – it was different for the tourists and that's

what they came to see. Now the tourists don't like it because the young people are also like tourists, and all the new buildings and cars and so on. It isn't different for the tourists now, so they don't like it. And it'll get bigger and bigger – with more tourism things, because the old people will die and the young are like tourists themselves, so it will become like Ürgüp. Göreme's gone, it's finished. I miss the old times when people rode donkeys to the garden.

This comparison with Ürgüp is frequently made by Göreme villagers, and is indicative of a desire to retain a sense of control over, at the least, the tourists/guests in their village, even if they do not have control over the shape of the future of tourism *per se*:

We are much better than Ürgüp, really, because Göreme is much better as a place, as a village, so travellers prefer to stay in Göreme and so we have a chance to make business with them.

The aestheticisation of the old village life by villagers is not only nostalgia for times past, and nor is it purely the adoption of tourist discourse in pursuit of economic gain, but it is also born out of a fear of moving towards a situation like that of Ürgüp where tourism business and development have got out of local control.

Concurrently, the Göreme villagers often feel restricted by the preservation orders regulating what they can and cannot do to their own houses.[6] In conversation with a carpet salesman, I was told:

Suppose you have a cave, and you have been living there many years, and your toilet is falling down, and you want to build a new one, and all those officials come to you and tell you cannot do it...And probably they are from Ankara or Nevşehir, and they are just sitting at their desk, and probably they have never been to Göreme. It's becoming a big problem. They say: 'This is the rule, it is forbidden, it is the law.' But it may not match with people's life.

He continued:

I wish the director of the National Park was someone from the region, who knows the region, not someone who is from somewhere else. I wish he lived here, that he was a man from the village, so that he could help the people. I wish the people could live in Göreme for longer, and their future generation. If they need new homes, it must be permitted by the Ministry or the director or whoever does that.

Villagers experience the anonymous direction of the external National Park and Ministry of Culture authorities as having direct control over significant aspects of their lives, including whether or not they may carry on living in their cave-houses and in the village. There is also an awareness among villagers that the aesthetic valuing and preservation rhetoric surrounding their cave-homes is connected with tourism and what tourists want to see. When I commented on the darkness of a young woman's cave-kitchen, for example, she said:

> Yes, but we are forbidden to make new windows or shelves or anything in the rock. Before it wasn't forbidden, but when tourists came here it became forbidden.

Another villager told me:

> In some ways we are lucky because tourism has brought work and money, but there are also many forbidden things, and expensive things. We can't do anything – we need permission for everything. Even on little alterations to our houses we must use the right stone and so on, and we can't build new caves for the animals.

Indeed, many villagers have been brought in front of the law as a consequence of alterations undertaken on their rock houses, many of which were carried out to convert the houses into tourist *pansiyons*. The building and alteration legislation hence manifests itself in the face of tourism as a series of double binds for villagers. First, villagers' experiences of a lack of choice and control over the homes that they live and work in are recognised as stemming in some way from tourism. However, because tourism brings chances of prosperity to the village, the villagers themselves also take on board these same values concerning the need for preservation and protection of the village. Tourism brings villagers the chance of financial gain, but at the same time brings the building regulations that may deny entrepreneurs permission to build or improve their tourism businesses.

While the tourism business in Göreme is largely locally owned, providing villagers with a significant amount of control not only over their livelihoods, but also over their relationships with tourists and over the tourists themselves, the representation and preservation rhetoric surrounding the village is externally and hierarchically imposed. Villagers have little say in the management of Göreme's cultural 'assets', such as in the zoning plan developed by the Ministry of Culture's Preservations Office in Nevşehir, which designates different areas of the village for tourism, new housing or preservation. They can thus feel a strong sense of injustice in the fact that they have restrictions placed on their ways of life.

It is important to remember, however, the heterogeneity of communities and that, as Stonich points out,

> Social heterogeneity (inequality, diversity of interests, latent as well as overt social structures within even the smallest settlements, the complexity of

local cultures, etc.) has immediate implications for successful community participation.

(Stonich 2000: 149)

In Göreme, while there might at times be a predominant relation of 'authority' versus 'villagers', preservation rhetoric expounded by external authorities is adopted by some villagers and may even be used by them against others. As I pointed out in Chapter 5, *pansiyon* owners may alert the authorities to some 'damaging' and illegal building work carried out by a neighbouring *pansiyon* in order to have a large fine or even prison sentence inflicted on their key competitor. In one particular case, a *pansiyon* owner was imprisoned for a year because he whitewashed over a faded Byzantine fresco in order to 'clean' a cave-room in his *pansiyon*. It was reportedly villagers who lived in houses surrounding the *pansiyon* who had alerted the authorities in the hope that the *pansiyon* might be closed down: they did not like the presence of a bar and *giaour* (infidel) activity so close to their homes. I was also told by some villager entrepreneurs that it was good that this man was put in prison because it served as a warning to others not to damage the historic value of the village. This is indicative of the growing discourse among villagers concerning the preservation of the old aspects of Göreme.

These complexities, both in the official rhetoric surrounding Göreme and in the way this rhetoric is filtered through local values, discourse and practice, clearly convey the tensions created by the conflicting sets of values regarding Göreme local identity and place. On one level, this hotbed of cultural negotia-tion is tempered and obscured by the villagers' more mundane concerns with the practicalities of their day-to-day lives. On another level, though, the theme of preservation and protection of the natural and traditional landscape of Göreme forms a highly visible seam that runs right through tourist, official and villager discourse concerning Göreme – and is hotly contested.

The inevitability of change

Cultural negotiations between tourists, villagers and the various authorities are clearly vibrant in their intensity, and social relationships among villagers, and also incomers, are alive with the push and pull of ties and contentions just as any rural village might be. This of course means that Göreme cannot remain in any sort of static state, protected from change and the perceived homogenising forces of modernity and, especially, tourism. Indeed, change is inevitable, and anyone who stays in the village for a prolonged length of time gets a sense of rapid change: marriages; deaths; a sense of the transition from a poor past to a hopefully prosperous future; changes in the season and the yearly opening, closing, reorganisation and changing in appearance of the many tourism busi-nesses.

The Mayor of Göreme, as the head of the municipality (*belediye*), is primarily responsible for building planning, and is therefore in a key position to effect change, at least to the physical appearance of the village. The man who was mayor for most of the 1990s, Mustafa Mızrak, carried out a great deal of work to 'beautify' the village centre by planting flowers and trees and extending and tidying the look of the bus station. In the early 1990s he also built the 'tourism centre' – the one described as 'hideous' in *The Good Tourist Guide to Turkey* (Wood and House 1993), quoted in Chapter 2 – accommodating tourist shops and an open waterpool area, and in 1997 he built a row of shops in the Göreme Open-Air Museum bus park in order to provide more permanent accommodation that the villagers with souvenir stalls there could rent from the municipality.

These developments promoted considerable debate among villagers, however, concerning what is 'best' for the village and for tourism. As a member of the Anavatan Partisi, a centre-right political party, Mızrak was voted in because of his keenness on promoting tourism in the village, as compared to the other choice at the time who was a member of the religious Refah (Welfare) Party and supposedly against tourism development and the infidel behaviour it brings. Indeed, Mızrak did a lot to promote and accommodate tourism development in the village, but he was also frequently accused of focusing too much on making the village centre more 'beautiful' for tourism, to the neglect of the residential areas. A case in point here concerns the improvements made to the central road running through the village that allowed traffic to move much quicker along that road than was previously possible. Also, because of the increasing number of large tour buses passing through, the road became extremely dangerous for villagers driving their donkeys and horse-carts. Consequently, donkeys and horse-drawn vehicles were supposedly banned from the main road, and so villagers are left to skirt around the centre using the badly conditioned dirt roads, to which – they complain bitterly – the municipality rarely pays any attention.[7] Villagers said they would have liked to see more work carried out on the residential streets to match the improvements made to the central area of the village.

Another change effected, in part while Mızrak was mayor, was a transformation in village marriage celebrations. Village weddings would usually take place over three days, with separate dancing parties (*düğün*) for men and women occurring in the courtyards of the respective families' homes. Marriage in urban Turkey, however, now takes a more 'Westernised' form, with a one-evening party in a 'wedding salon' of a hotel, attended by men and women, and the bride and groom, together.[8] During the first year of my fieldwork the son of the mayor was married and, as a usual village wedding was inappropriate for such a politically conspicuous occasion, Mızrak arranged a modern-style marriage party on the swimming pool terrace of the 'Turist Hotel'.[9]

We saw in Chapter 5 that because the 'event-rich' quality of village life renders all change highly visible, villagers tend constantly to copy each others' patterns of behaviour in an effort to remain equal. While such changes may clash with other cultural values, such as the codes of shame and honour that surround women's movement in the village, the 'copying' can snowball so rapidly that changes quickly become normalised into a new *adet* (custom). Schiffauer notes similar processes of change in the Turkish villages he studied:

'Because status is involved, imperatives are set, which cannot easily be changed even if they seem to be irrational' (Schiffauer 1993: 75). Following the wedding of Mızrak's son, numerous similar wedding parties were held in the same hotel, even though it is difficult, and at the least uncomfortable, for village women to attend such an event (see Chapter 4).

Another change that has snowballed in this way is the sending of children to private schools outside the village where it is believed they will be better educated in English and computer skills. This trend was begun by a few tourism entrepreneurs in the village who, on the one hand, became able to afford such schools for their children and, on the other hand, believed that this education would equip their children better for tourism work, or some other non-agricultural work, in the future. As girls as well as boys are beginning to be sent to these schools, the trend might have a profound influence on village gender roles and relations in the future. A further change likely to influence gender and economic relations into the future is that mentioned in Chapter 4 whereby, because of the income from tourism, many villagers are now able to relax their agricultural efforts and instead buy goods from the market. We are reminded once again of the argument I overheard between Abbas and his wife about whether to burn the wheat field and be done with it.

The aspect of inevitable change that is particularly pertinent to attempts to preserve the current (or past?) state of the village is the natural erosion of the 'fairy chimneys' and cave-houses, which continuously works to alter and remould the Göreme landscape. The restrictions determined by the Ministry of Culture Preservations Committee forbid any new digging into the existing caves in the Göreme area. Yet it is precisely such continuous new digging that has allowed populations to be resident here for centuries. As older chimneys and caves collapse, new cave-homes are dug out and that is the process whereby the village of Göreme today has come to be situated two kilometres from the original monastic site of Göreme – now the Open-Air Museum. Though technology is currently being used to rebuild some fairy chimneys containing the best Byzantine frescoes, this restoration process is too costly for extensive restoration throughout the entire area. It could be argued, then, that by forbidding further digging out of new cave-homes, the preservation laws are actually preventing the continuation of Göreme, the place, because the village will gradually, and inevitably, be eroded away. A villager entrepreneur made this point when he was complaining to me about the preservation restrictions in the area:

> It's forbidden here and forbidden there. The thing is that if the carving of the rock was forbidden in Christian times, we would never have those churches! When you go to the valleys you can see big rocks which are not touched, which have never been carved. So people should be allowed to carve them, to use them as storage at least, or later on it could be used for other purposes, so that maybe a hundred years later we can leave something for the future people – so they can come and visit our homes.

Plate 8.2 Erosion of the fairy chimneys: 'Disneyland tells us that technology can give us more reality than nature can' (Eco 1986: 44). Funding from the Turkish Ministry of Culture and UNESCO pays towards these efforts to strengthen and rebuild the fairy chimneys that house the best examples of Byzantine frescoes in the Göreme Open-Air Museum.

Both physical and social changes, short-term and long-term, are rife in Göreme, and such change is inevitable. Therefore, any efforts to somehow freeze the village in some static form, in order that it will continue to remind tourists of 'the past' well into the future, are surely hopeless.

Emergent culture: the Flintstones of the future

Of course, tourism development and villagers' opening and working in *pansiyons*, restaurants and tour agencies are all significant recent changes. Moreover, while these businesses are places of work for the Göreme men, they are also *liminal* zones that provide new contexts in which the men are relatively free from the village way of life and village rules. This idea was expressed in an interview with a village entrepreneur when asked how he felt when in his *pansiyon*:

> It's like a free zone – I can't walk in shorts on the street but I can here. So I must be careful to change when I go back into Göreme. I can't walk with a girlfriend through Göreme holding hands, so I'm happier here – I feel more free.

This sense of being apart from village life while in the tourist 'enclaves' is compounded by the fact that, during the summer tourist season at least, the men work very long hours and often sleep in their businesses. During the summer, they may rarely go home, even if their home is only some metres away from their business. They also become neglectful of village events such as weddings and funerals. Some village men even placed their experiences of working in tourism in the village together with the experiences of villagers who have migrated out of the village to work in the cities or even in northern Europe, the key point of similarity in the two experiences being the difficulty in being re-accepted on 'return' to village life in the winter.

There may be negative aspects in their experience of being 'outside' of the village while working in tourism, but the men nonetheless invariably enjoy being able to join in the 'play' of tourism. As I heard one *pansiyon* owner saying to his guests, 'You are on holiday now, but we are always on holiday here.' The men are able, to some extent at least, to join in with the fun and fantasy of the tourist experience, and in Göreme this includes the playful performance of being troglodytes in a cave-land fantasy. Together with the tourists, therefore, the Göreme men play and experiment with their own identities, engaging in an ironic play on touristic representations concerning their 'cavey' identity.

Indeed, as I described at the beginning of the introduction chapter, one of the young Göreme men has named his *pansiyon* 'Flintstones Pansiyon' after the cartoon comedy that anachronistically depicts cave-dwelling people living out a modern lifestyle in a prehistoric cave-land environment. *The Flintstones* is shown in Turkey on satellite TV and, based on this, other businesses in the village have names such as 'Bedrock Travel Agency', named after the town where *The Flintstones* cartoon is set; for some years now, tourists have been hearing about the 'Flintstones Cave Bar' long before they arrive in the village. Many other businesses follow the same theme, with names such as 'Troglodyte Pansiyon' and 'Rock Valley Pansiyon'. The owner of the 'Flintstones Cave Bar' told me the following about how his 'play' resulted in his use of a Fred Flintstone characterisation:

> A few years ago in this *pansiyon*, there were four Aussie girls sleeping in the cave in the fairy chimney – I was born in there – and I had to give them an early morning call and, just for a joke, I shouted 'Wilma wake up!', and they said 'We're coming Fred!', and so they gave me that nickname, of Fred. And they sent other tourists here later, telling them to go to Fred's place. It began like that, and they liked it, and they sent me socks of Fred Flintstones, alarm clock of Fred,...and then I decided to call the bar 'Flintstones Bar'. It's a really good name, because it's in the rocks, a real cave, like the Flintstones, it's the Flintstones bar – the Flintstones movie – it's so famous in the world, but it's fun. Of course, people come here and I invite them to come for a drink in the cave-bar, and they say 'This is fantastic, who did this?' And I say 'I did that.'

Göreme has consequently become a fantasy-land of caves and troglodytes, a sort of Disneyesque Flintstones World where tourists can stay in a cave-room in the 'Flintstones', 'Peri' ('Fairy' – as in *peribacaslar*, meaning 'fairy chimney') or 'Rock Valley' *pansiyons* and book a day tour of the area entitled 'Mystic Tour', 'Fairy Tour' or 'Dream Tour', chosen from a cartoonified regional map in the offices of the 'Magic Valley' or 'Bedrock' travel agencies. At night, tourists can go to the 'Flintstones Cave Bar' or 'The Escape Cave Bar and Disco', which is 'set in gigantic medieval donkey stables'. There, they can watch a 'traditional Turkish belly dance act' and 'dance to the latest in dance music'[10] together with some of the local troglodytes, who might tell them about how they were born and brought up in a cave and how the 'fairy chimneys' come alive at night.

The villager quoted above often introduces himself to newly arrived tourists as Fred, a local caveman. He points to a cave and tells tourists that he was really born in a cave right here. He also calls his dog 'Dino' after the Flintstones' pet dinosaur, and collects Flintstones paraphernalia to decorate the office of his *pansiyon*. Fred was indeed born in a cave-room of the house that has now been converted into his *pansiyon*. However, he might be considered today as a caveman in something of a *post*-modern sense, rather than in the pre-modern sense conjured up in much of the tourism promotion literature and preservation rhetoric concerning Göreme. In the first place, the link with the famous American comedy cartoon is in itself an indication that Göreme village is not so pre-modern as to not be linked up with the global network of terrestrial and even satellite TV. Furthermore, it is the actual way that the imagery is manifested

Plate 8.3 Flintstones imagery used in Göreme's tourism.

in the villagers' performances, and the way it is negotiated together *with* the tourists, that becomes significant here. For the Flintstones imagery is generated largely through the idiom of irony that arises in the interactions between the tourists and the villagers, and it is this sense of irony that enables the men, through their performances, to both play to and at the same time resist the 'pre-modern' caveman representations of themselves in tourist discourse and official rhetoric.

By performing this caveman identity for tourists, the local men are bringing themselves into the foreground and acknowledging the role that this particular aspect of their 'traditional' identity plays in Göreme's tourism. Interestingly, the owner of another *pansiyon*, called the 'Flintstones Pansiyon', is the son of the family who take tourist groups into their cave-house as described in the scenario in Chapter 6. He has, of course, seen only too often the images and representations that the guides who bring groups of tourists to his parents' home use to paint imaginary pictures of his family. Indeed, all of the tourism entrepreneurs are well aware of the importance that tourism discourse places on the unique and natural qualities of this cave-village. As we have seen above, it is particularly through the official rhetoric concerning the preservation of the village that villagers have come to realise the value of their seemingly traditional 'cave-life' as an important tourism asset.

Not being prepared, however, to accept in their interactions with tourists a troglodyte or peasant identity that merges with a sense of static tradition and even *backwardness*, villagers present themselves and Göreme as a sort of comic fiction. In an interview with Fred, for example, he told me about a time he got into an argument with a man in Istanbul. The man said to him: 'What's wrong with you, were you born on a mountain?' Fred answered: 'No, I was born in a cave!' It is precisely through this same sense of fun and irony in their interactions with tourists that the men are able to re-negotiate and divert some of the representations placed on them by tourist discourse and the complex multiplicity of authorities and tourism bodies who assert a need to preserve their cave identity. Indeed, in some parts of the village the more traditional cave-life does still exist, and there tourists may wander and have their more serious experiences with the 'authentically social'. In the tourist realm too, the villagers are retaining their 'cavey' identity, but here they are doing so together *with* the tourists – the same tourists who may at times be more serious – in an ironic twist through which they are able to bring their ascribed pre-modern identity into the realm of the (post-)modern.

Experiencing the hypo-reality of Göreme

For the tourists, the Flintstones characterisation in Göreme is experienced in a variety of ways. The four Australian girls who Fred talked about in the quote above clearly enjoyed staying in the cave-room and playing the Flintstones game, while others, as we saw from some of the tourist vignettes in Chapter 3, make a clearer distinction in their constructions of the village between the tourism realm, which they see as servicing them in modern comfort, and the 'traditional' realm, where they can experience the 'authentically social'. Others still, feel that the Flintstones characterisation is so obviously contrived for tourism that it, to their annoyance, blurs their experience of being in a 'natural' cave-place.

An example of this last view came from the Belgian tourist quoted at the beginning of this chapter. This quote epitomises the hegemonic discourse of cultural preservation and the desire for places like Göreme to be fixed in time. This was also expressed in the Belgian's imagining that in ten years' time Göreme is likely to be too touristy for her to be able to enjoy it – this is seen as manifested in touristic signboards and neon. As I have said, tourists frequently express this trope of the vanishing primitive whereby 'for generations of tourists,...primitive peoples have always been seen as on the edge of change, to be experienced or described before they disappear' (Bruner 1991: 243). The important point here, however, is that although Göreme seems always to be on the verge of tipping the balance towards becoming over-touristified, it also seems to stay on that verge 'for generations'. Picard also makes this point in relation to Bali: 'to each new generation of visitors, Bali seems to be on the brink of ruin, holding out by a reprieve of the good fairies' (Picard 1996: 92).

The recent Internet explosion in Göreme nicely illustrates this point. Although the many cybercafés that were springing up in the late 1990s might initially jar with tourists' expectations of this rural cave-village, they are frequently used by the tourists and seemed even to add a quirky element to tourists' perceptions of the village. Hence, while the discourse of preservation of tradition is very much at work in Göreme, the tourists seem to be adapting to fit the place just as much as the place is adapting to fit them. In other words, as the performances of the villagers in their interactions with and production of services for tourists become increasingly multifaceted regarding their presentation of a 'traditional' identity, the desires and experiences of the tourists simultaneously increase in their ambivalence concerning the 'traditional' in the village, as least in its static sense. This is because, regardless of whether the presentations and performances of the village and the villagers are perceived as authentically traditional by the tourists, the touristic encounters in themselves meet with the tourists' quests for the 'authentically social' precisely because they are not blatantly staged, and so in themselves provide the cultural difference that the tourists are seeking.

Göreme village, both in its division into two different spheres and in the performances of the Flintstones characterisation, appeals to the ludic tendencies that tourists have alongside their desires to experience the village as a real 'cavey' place. As the village and the villagers change to become increasingly touristed, so tourists seem to be prepared to accept it, as long as they are having fun and as long as they get a sense that the performances are embedded at least partly in the 'real'. This is illustrated in the following short anecdote told by the owner of the Flintstones Pansiyon:

> I was talking to some tourists who I met on the street in the centre, and I said, 'Let's go to the Flintstones Bar', and they said, 'Oh no, it's too modern.' I said, 'Have you been there yet?' – they thought the name was too modern, too artificial, but they hadn't been there. But when they went to the place, they had a different idea, because – it is in a cave! (he laughed) and very natural!

Indeed, the idea that places might be 'too modern, too artificial' is the idea that there are two broadly opposed types of tourism and tourist places: the vulgar, fun-loving type taking place in contrived sites, and the authentic type occurring in real and natural settings.[11] Dichotomies, or at best continuums, are constructed between such concepts as travellers and tourists, romantic and crass, authentic and contrived, real and fake. While it is generally assumed that tourists and tourist places are either one or the other, it seems from close analysis of the tourism in Göreme that the two might be combined. This is because Göreme the place not only seems to appeal to various touristic tendencies at the same time, but it also seems to collapse tourists' experiences of 'real' and 'fake' into one.

Indeed, it has been widely noted in recent academic discussions on tourism that contemporary tourist experience is characterised by a suspension of 'the saliency of the boundaries between...fact and fiction, reality, reconstruction and fantasy' (Cohen 1995: 20). This point, however, is most often discussed in reference to the proliferation of contrived or simulated tourist sites, such as theme parks, and the perception, highlighted by Eco (1986), of the fake as more real than the real. Göreme, on the other hand, is a natural place with a culture embedded in local tradition, and is importantly perceived as such by tourists. And yet, because of the 'fantastic' landscape there, together with the Flintstones characterisations played out in the touristic centre, the village can be imagined and experienced, alongside the real and traditional, as somehow artificial. The landscape of Cappadocia is naturally formed and yet is perceived by tourists as being so weird that it is often described as 'Disneyesque'. It is like a 'huge adventure playground', tourists say, a sort of 'moon world', 'like a different planet', 'it's unique, visually stunning, weird, the most abstract place I've ever been to'.[12] As one tourist said, as she stood looking at the weird rock shapes pitted with steps and doorways carved through centuries of real living,

> I'm having a hard time believing this is real. I guess I've been influenced too much by Disney World where they make things like this out of poured concrete.

The suspension here of the boundaries between the modes of experience marked by reality and fantasy are striking.

A further example concerns a young American tourist I chatted with in the village. He viewed his travel as an escape from the rat-race back home and his 'unethical' job as a maker of videos and computer games, but found himself feeling keen to return to the USA so that he could set about making a computer game of Cappadocia. The game would feature moonscape valleys and underground networks of caves and tunnels, and the player would enact an early group of Christians fighting off the attacking Hittite or Persian armies. This man's ideas are indicative of a developing global culture of tourism that accepts anything or any place being produced and reproduced, moved and recontextualised in any place whatsoever. Usually referred to as post-modern, this process marks the proliferation and increased consumption of experiences characterised by:

> stylistic eclecticism, sign-play,...depthlessness, pastiche, simulation, hyper-reality, immediacy, a melange of fiction and strange values...[and] the loss of a sense of the reality of history and tradition.
>
> (Featherstone 1991: 76)

In an age of simulation and a world where one moves freely and easily between the real and contrived, there is instilled a belief that anything is capable of being reproduced. Therefore, 'the "completely real" becomes identified with the "completely fake"' (Eco 1986: 7), and the ability to distinguish between the two is lost. Theme parks, then, are not only created worlds of fantasy, but places of hyper-reality where 'absolute unreality is offered as real presence' (ibid.). In theme parks, latex crocodiles in the Amazon jungle are more lively than the real thing, and two-dimensional pictures can come to life. The point for Eco is that simulation appears as more real than the (really) real.

My point concerning Göreme is a continuation of Eco's argument but in the converse: If simulations are experienced within the post-modern ethos to be more real than the real, then the other side of that coin must be that the real appears to be more fake than the really fake. Göreme is not a theme park, a fantasy-land created commercially for tourists' entertainment and recreation. Yet tourists frequently make comments such as, 'This place is like a cross between a boardgame and a fairy-tale – it's unreal!' Some tourism, then, may become travel in *hypo*-reality (a reduced sense of reality), and that, I suggest, is what is happening in Göreme when tourists have trouble deciding whether it is a 'real' place that they are in or a Disney World created out of poured concrete. This certainly seems to be the idea expressed by an English tourist in the following:

> It's pretty cheesy, you can't take it seriously. I said when I came here a few years ago that it was like a film-set, but now it's more like a theme park – but it's really nice at the same time – but it's really unreal – with the fairy chimneys like they were made out of polystyrene. They're mad formations that you wouldn't think could be formed, and with all the Flintstones stuff, like Bedrock this, and Fairy Tour that, it takes out the spirit of it being a real place. It's more artificial – it's a Cappadocia theme park.

Furthermore, it is clear from these comments that the experiencing of a sort of hypo-reality in Göreme does not necessarily interfere with tourists' enjoyment of the place. Although tourists may come to the village initially with a desire to experience the traditional in a more natural sense – and while a few tourists do experience tension in the juxtaposition of new and old, real and contrived – most seem quite easily to be able to suspend the importance of such serious matters as long as they are having fun. An English tourist who was riding a motorbike around Turkey described his drive up from the south coast to Göreme like this:

> I was coming up onto the Anatolian plain, and as I was passing villages I felt as though I was going back in time. That was until I came over the hill and into the Göreme valley, and realised that I had arrived in – Blackpool!

This tourist proceeded to really enjoy his stay in the village, having adventures exploring and clambering in the valleys and caves, and ended up staying quite a few more days than he had originally intended.

The continuation of Göreme as a tourist site?

While Göreme village is represented in tourism discourses as a traditional village of cave-dwelling peasants that should be gazed upon and preserved under a shroud of authenticity and preservation, in the tourist arena within the village, tourism services display names and imagery that engage tourists in a sense of fun and irony by presenting Göreme as a Flintstones fantasy-land. It is precisely through this sense of irony that the Flintstones characterisation played out in touristic performances simultaneously feeds and diverts certain tourist representations, or the primary tourist gaze, concerning the Göreme people and place. So, although the Flintstones characterisation works to convey the 'other' identity that tourists expect, it also goes some way towards defying the various authorities that have a hand in managing tourism in the area. The local men are reinventing, and even re-contriving, their 'cave' identity; in doing so, they are resisting something of the 'traditional' and 'backward' identities placed on them, and thereby some of the limitations and frustrations they face under the weight of the hegemonic discourses concerning the value and preservation of *their* place and life. As Chambers has argued, any attempt at cultural preservation cannot be authentic unless it is done through a process of participatory, rather than hierarchical, decision-making. Therefore,

> a community that has the ability to decide to tear down all its historic buildings in order to construct a golf course for tourists is more authentic than is another community that has been prohibited by higher authorities from doing the same thing in order to preserve the integrity of its past.
>
> (Chambers 2000: 99)

Importantly, also, most of the tourists who stay in Göreme experience the 'new' cultural differences in this way, enjoying the hypo-reality that has emerged through Göreme culture's interaction with tourism. This links directly with the issues raised at the beginning of the book regarding the sustainability of Göreme as a 'cultural tourism' site and whether or not Göreme might become so de-traditionalised and so over-touristified that tourists will no longer have any desire to go there. It has been shown that tourists can have multiple and negotiable identities and experiences, and also that, in Göreme, they seek the 'authentically social' in a variety of ways because they engage in a second, interactive, gaze as well as the primary gaze of tourist representations. They not only seek some sense of authenticity embedded within the places they visit, giving rise to the preservation rhetoric outlined above, but they also desire interactions with local people and places in ways that are not planned or staged for them. The importance of this interactive gaze is two-fold. First, it suggests that for contemporary tourists it is the unexpected in their interactions with local people and places that holds the key to what has often been referred to as 'authenticity' in tourist experience. Second, it allows us to see where tourists' quests and experiences are open to negotiation together *with* the environments and people they interact with during their trip.

Understanding this point allows us to step out of the corner we are forced into when we associate cultural difference with some imagined past, and when we view difference as a property fixed within local people and places. The problem, as mentioned above, is with the idea that there are two opposed ideal types of tourist places and experiences that can be characterised by such concepts as natural and contrived, real and fake. Many discussions of tourism have asserted the view that places, together with tourist experiences of those places, can only be one or the other: authentic or contrived, non-touristic or touristic. However, in reality, where ideas of place, otherness and ourselves are influenced through tourism in highly complex ways, such dichotomies do not necessarily hold: the tourists in Göreme delight in adventure and the unexpected in their touristic interactions as much as they delight in experiencing a 'real' cave-life in the village; they say they are not tourists, but then they buy a day tour of the sites of Cappadocia; they enjoy the chance to wander around the back streets of the village where the 'traditional' activities are performed, but they also relish the fun of the bars and discos in the evenings in Göreme's tourism realm.

A key point here is the irony that these tourists engage in, together with the villagers with whom they interact, to enable them to cope with the paradoxes and contradictions inherent in their condition and in what they are doing. In Chapter 3 it was seen how tourists constantly play with and between various layers and levels of irony so as to negotiate their problematic tourist versus non-tourist identity. To miss the importance of this irony is to fix tourists and tourist places into the dichotomies outlined above, and this fixing in turn renders the chances of continued tourism quite hopeless in places like Göreme, precisely because it presents the need for preservation in the face of continuous and inevitable change and development.

Indeed, the landscape of Göreme is particularly conducive to this state of ironic play, because besides its being associated with a sense of the authentic and the past, it is also perceived and enjoyed by tourists as a cavey 'adventure play-ground'. Besides the clambering around in the tunnels and caves in the Göreme valleys during the daytime, tourists can spend the evenings in cave-bars and restaurants, and then at night they can even sleep in their very own cave in the *pansiyons*. Tourists can enjoy adventure and play in Göreme, not only in the fact that they are in a cave-village, but also in imagining that they themselves, for a while at least, are cave-people too.[13]

Furthermore, the temporary nature of the tourist's stay means that the tourist experience of a place and its people has a certain 'virtuality' about it: it is very much a surface experience and never really achieves a sense of reaching the core of the place. Indeed, if ever tourists do begin to get a sense that they are becoming more deeply embedded in the core of the place – by being invited to stay in local homes, for example – they often feel claustrophobic and have a desire to pull back out of the situation and place. This is an inevitability set by the laws of hospitality discussed in Chapter 6. While, on the one hand, tourists want to have close unmediated interaction with their villager hosts, the tourist condition is simultaneously characterised by wanting to be on the edge of places and peoples where it is possible to experience this sense of freedom and play. Furthermore, if tourists are on the edge of the places they visit, then that is surely where tourism workers must come to meet them. To a large extent, then, interactions between tourists and villagers take place within this context of limi-nality and play.

We should be careful here, though, not to get too carried away with the 'play' in tourism. As Selwyn remarks, if we lose sight of the distinctions between tourist fantasy and the socio-political economy of the tourism processes, 'there may be no way out of an eventual wholesale Disneyfication of one part of the world built on the wasteland of the other' (Selwyn 1996: 30). Indeed, the tourists' liminal experiences of fun in this Disneyesque fairy-chimney land are only temporary because in the end the tourist always goes home. For the local villagers, on the other hand, this fantasy-land is home – so where, we might ask, is their reality?

The tourists see Göreme and Cappadocia as a magical land of fairy chimneys and cave-dwellers, and a monumental site of Byzantine history. Many tourists see Göreme as a place in which they can experience the 'authentically social'; as a place in which life is aesthetically natural and 'wholesome'; a place that somehow got stuck in the past. Concurrently, tourists seek the company of like-minded backpackers in Göreme's *pansiyons*, bars and discos, and for them the moon-like landscape is an appropriately 'other' environment in which to have fun and adventure: it is a Flintstones fantasy-land. Many of the Göremeli men, too, enjoy their new lives with tourism, and they fight the pulls of the social and moral order of their lives in the village in order to participate in the freedom and fun of the tourism realm. For the villagers also though – both men and women – the village is home, land for gardening and, more recently, a place of business and economic prospects. Indeed, beyond concerns about the weight of preservation rhetoric on villagers' activities, and beyond the repercussions such hegemonic values have for their identity, the villagers' 'reality' is necessarily structured by their more practical desires to improve their economic situation and to have some control over their lives in the tourism context.

This multiplicity of meanings and values associated with the local landscape and culture is bound to give rise to tensions and clashes, and we have seen throughout this book that these are plentiful in Göreme. We saw in Chapter 6 how hospitality, which lies at the centre of the villagers' identity, is gradually being abused and eroded so that they are losing their sense of place and control in the tourism context. The rapid opening of small businesses in recent years has also led to a sharp increase in aggressive and violent competition between villagers, and the increasing numbers of 'outsiders', both Turks and tourists, trying to carve out their own fortune from tourism business adds to the tensions. Likewise, the presence of tourists has given rise to new forms of gender behaviour and relations that are often seen as problematic by villagers. Also, the increasing social separation of the tourism realm from the 'back' realm and the households leads to a sense that the young men, even without their actual outward migration, have somehow left the village. The situation, therefore, is far from easy.

Concurrently, however, there is no doubt that most of the villagers also see this level and type of tourism business, which they are able to fully participate in, as a blessing because of the economic opportunities it has brought them. Moreover, by providing these new chances of prosperity within their home village, tourism has lessened the need for young men to migrate out to seek work; thus the village, while becoming divided into two separate realms, has in many ways stayed more intact and full of hope than it might otherwise have done. In this, their ability to position themselves as hosts to their tourist guests has largely enabled the Göreme villagers to have a say in negotiating their interactions with tourists, and this is a negotiation not only of the 'traditional' identity of the villagers and their way of life in the village, but of the tourists' quests and experiences in themselves, so that neither the tourists nor the tourist site should be viewed as static and fixed. It is precisely through this negotiation, rather than through 'preservation', that Göreme may continue as a tourist site into the future.

A poignant illustration of the tensions and negotiations embedded in 'place' with which to end this chapter, comes from a situation I observed at a circumcision party that was held in a *pansiyon* converted from an old cave-home. Appropriate to village custom, the men and women had separate seating areas and, while the men took their place in the more public area of the *pansiyon* courtyard, the women were sent downstairs to a cave cellar that was otherwise used as a cave-bar in the running of the *pansiyon*. I joined the women, wrapped in their usual layers of clothing and headscarves, in the cellar and, as I sat among them, I realised that the walls of the cave had been painted with dark and harrowing images of skulls, cross-bones and naked women hung from crosses. At first we sat more or less in silence eating the festive meal, but thankfully the tension was then broken by an old woman cracking a joke about how we were made to eat our feast among the naked ladies. What these women must have thought about what goes on in the minds of tourists, and indeed in the minds of their husbands and sons, if they paint tourist bars like that, is anybody's guess.

An important point to remember, though, is how quick these women were to put an ironic slant on the situation in order to cope with their initial shock, just as the villager entrepreneurs have been quick to twist their caveman identity around to their own advantage in the realm of tourism business. As we have seen, the tourists too are largely able to cope with and derive fun from the Flintstones-land in which they find themselves. As long as certain conditions prevail, conditions that allow both tourists and their hosts to play an active role in negotiating and determining their identities and experiences, the 'new' that emerges through tourist meetings can be accepted, so that tourists and troglodytes will continue to meet in Göreme for some considerable time more.

Notes

1   A copy of this report, prepared for the Turkish government in the late 1960s (exact year not stated), was located in the library of the Turkish Ministry of Tourism, Ankara.
2   Although, besides the National Parks Directorate (under the state Forestry Department) and the Preservations Committee, the Ministry of Tourism and the Göreme Municipality Office also have a stake in preservation in Göreme, and it is neither surprising nor untypical that these political bodies each tend to work to their own agenda. Hence, the use of political affiliations, social connections and bribes have reportedly played a large part in what building and business has or has not been permitted.
3   See, for example, Allcock (1995), Drost (1996), Herzfeld (1991), Hollinshead (1997) and Urry (1990).
4   That both the Save Göreme Committee instigator and the *Lonely Planet* writer are from northern Europe seems quite significant here, given that it was Western travel writers and US planners who initially romanticised the 'picturesque' rural cave-life in the village.
5   It is only recently that non-Turkish nationals can legally own property in the country, and increasing numbers of foreigners are now beginning to buy old cave properties in the back streets of the village to restore them as houses and businesses. Villagers are keen to sell houses to 'tourists' to get a high price. For discussion of similar processes occurring elsewhere, see Waldren's accounts of 'outsiders' moving into the old village of Deia in Mallorca (Waldren 1996, 1997).
6   Similar processes occurring in rural Greece are described by Herzfeld (1991) and Williams and Papamichael (1995).
7   Towards the end of the 1990s, villagers often complained about Mızrak's irregular practices regarding business licence demands and also building against regulations. Prior to the elections in 1999, they asked Fervzi Gunal, a respected Göremeli man with more socialist leanings who was residing in Ankara, to return to the village to stand for mayor. Gunal complied with the villagers' wishes and at the time of writing still stood as mayor in Göreme.
8   For more detailed accounts of wedding celebrations in Turkey, see Delaney (1991) and Magnarella (1974).

9 This particular hotel is the subject of much controversy in the village. It evidently defies the preservation and building regulations, especially as some whole fairy chimneys were knocked down to make way for its construction. As it is joint-owned by the Göreme Municipality Office and a governmental hotel chain, villagers frequently denounce the dubious way that permission was obtained for its construction.

10 These phrases are quoted from an Escape Cave Bar advertising leaflet.

11 For discussion on these two types of tourism, see Cohen (1995), Munt (1994a) and Urry (1990); see also Selwyn (1994, 1996).

12 This kind of 'fantasy' talk is argued by Dann (1996) to link the (adult) tourist with child-like characteristics.

13 Again, the idea of the playful tourist has been developed in particular by Dann (1996) in his assertions that tourists can be child-like. This idea also links with the concept of liminality in the ritual process (Turner 1969) as it is often related to tourism, whereby the usual order of things is removed and rules are reversed (Graburn 1983; Jafari 1987).

9   Conclusion

Writing tourists into destinations

This book has looked at how the variety of people involved in Göreme's tourism deal with and negotiate the new cultural identities, practices and relationships that have inevitably emerged as a result of that tourism. Juxtaposed with the gendered separation of lives and roles in the village, the tensions seen to exist between 'tradition' and 'tourism' have created the emergence of two distinct realms in Göreme for both tourists and villagers. While the back streets of the village are shrouded in preservation rhetoric, the central area comes closer every year to resembling a Flintstones theme park. Each of these two realms constantly tugs on the strings attached to the other: men's fun and sexual relations with tourists are checked and inhibited by their moral ties with their families in their 'home' lives, and likewise the limits of the codes of honour and shame concerning local women are stretched when women attend a wedding party by the pool in the 'Turist Hotel'.

We have seen throughout this book the ways in which the people of Göreme engage and are continually adapting with the socio-economic opportunities that accompany tourism development, engagements and adaptations that are necessary to 'host' the tourist visitors within their village. This has meant the villagers' gradual learning and adoption of entrepreneurial behaviour, and also the adapting of their codes of hospitality in order to accommodate their tourist 'guests'. This has involved an increasing sense of risk for villagers of their hospitality being abused by tourists, and so the giving of hospitality may be problematic and simultaneously, at times, confusing for tourists to receive. However, we have also seen how the idiom of the host–guest relationship may be used to redress the inequitable potential of tourist behaviours and representations, providing villagers with a sense of control over the tourists in their village, while providing tourists with the interactive and serendipitous experiences they desire. The exchange of hospitality and having unstaged interactions becomes the 'authentically social' in tourists' experiences, rather than that 'authentic' coming from images of a people who are left timeless in a pre-modern state.

While the initial quests and desires of cultural package tour tourists and independent tourists are not necessarily different in that they both follow the similar images and myths presented to them in tourist representations, what does differ is the degree to which their experiences while touring are structured and mediated by exogenous travel agencies and guides, and hence the degree and quality of their interactions with the locales they visit. Of course, there is no doubt that package group tourists can also have, and enjoy having, serendipitous experiences, but independent tourists actively seek and expect a close level of interaction with local people, and they therefore in turn allow those local people, in some degree at least, to play at being 'host' and to have some determination over the ways in which tourists view and experience them. In the 'back' realm of Göreme, where tourists seek the 'traditional', relations are negotiated largely through the roles and associated conditions of hosts and guests. In the tourism realm, where tourists seek fun and entertainment, hospitality again guides interactions, but negotiations between the tourists and troglodytes here also take place largely through an idiom of irony played out in this mutual liminal zone. When local people are in a position to demand relations of equality and respect with tourists, they can negotiate an identity for themselves that is more suitable to their inevitably 'touristic' situation than the static identities that are often placed on them by official rhetoric asserting the need for their preservation.

This is precisely why there is a need to include ethnography of tourists and their interactions in our analyses of tourism destinations and communities. Indeed, the words with which Selwyn concludes Boissevain's edited volume *Coping With Tourists* are:

> people in various parts of Europe have developed an extensive range of imaginative responses to tourism. People are, indeed, 'coping'. However, Boissevain's observation that they are 'coping *so far*', suggests that anthropologists and interested others need to consider the conditions that need to be met for local people to continue 'coping' and also those that might prevent them from so doing.
>
> (Boissevain 1996b: 253)

Through looking in this book at the interactions between the tourists and their hosts in Göreme, it has become clear that the 'conditions that need to be met' for people to cope are those conditions that allow for and play to the second, or interactive, tourist gaze. It is that interactive gaze that allows for a more dynamic notion of sustainability, a sustainability that accommodates social and environmental change, and recognises and accepts the new cultural forms that emerge through tourism.

The tourists in Göreme (and perhaps most others) try to get beyond the primary tourist gaze of stereotypical representations of extraordinary landscapes and 'other' cultures in order to individuate their experience and identity. One of the key ways they do that is through the purposeful avoidance of controlled and predictable situations: 'having no idea who I'm going to meet that day, or what I'm going to experience.' It is that element of surprise that is the essential element in the writing of experience and identity for tourists who manage to go beyond the primary tourist gaze. Serendipity, in other words, allows them to create the appropriate identity for themselves precisely by letting happenstance individuate their experience. Serendipitous events and interactions then become woven into and manifested in travellers' personal narratives, giving shape simultaneously to both their experience and their identity. Furthermore, because the world in which tourists move is therefore also active in writing their experience, that tourist identity is always under negotiation with other tourists and travellers, and also with the people and places they visit. An important part of the adventure, in other words, is letting the locale 'speak to' them.

This is where the institutionalised tourist gaze, which is a highly visual gaze, does not go far enough in explaining tourists' actual quests and experiences. Following that gaze, Urry has suggested that tourists may just as well be tourists at home these days, since it is possible to 'see many of the typical objects of the tourist gaze...in one's own living room, at the flick of a switch' (Urry 1990: 100). Urry is following Feifer here and her suggestion that, because of the media explosion, 'the passive functions of tourism (i.e. seeing) can be performed right at home, with video, books, records, TV' (Feifer 1985: 269). However, Urry presents this idea more strongly than Feifer intended it, I believe, precisely because of his over-emphasis on the primary gaze, on tourist representations and on *seeing*. Feifer rightfully recognises in the 'post-tourist' the validity of experience that goes beyond the purposeful gaze:

> She travelled at an unhurried pace, and there were just a few things she wanted to see: the maritime museum at Greenwich in England, the Norwegian fjords, and the Greek islands of Santorini; but she was looking forward, in a non-specific way, to whatever might lie in the way between them. It did not really matter what – it was a kind of random sample; not-seeking was a good way to find things of interest.
>
> (Feifer 1985: 261)

As noted earlier, MacCannell has also recently identified the meandering tourist who is interested most of all in that which is unexpected. MacCannell uses as his example Stendhal's *Memoirs of a Tourist* (1962), in which the tourist, Mr L——, travels primarily 'in order *to have something new to say*', and thus 'abjures a touristic sense of the extraordinary in favour of the unexpected' (MacCannell 2001: 32–3).

When recognising the second gaze, it becomes clear that to merely 'look' at places and peoples on television could not be altogether fulfilling to tourist experience because of the important part that interaction and happenstance play in the individualising of experience and providing 'something new to say'. A large part of tourists' quest is adventure, and that adventure includes unscripted interaction with the people and places they visit. If any experience or encounter is already decided on, packaged or staged as a tourist event, then they lose their own sense of framing and subjectivity and their interest is lost. And this was precisely the mistake made by the guide of the group tour of a cave-house in the scenario described in Chapter 6. The tourists were provided with the full picture of whole and harmonious troglodyte life in an effort to ensure a satisfactory experience of the 'authentically social' in their visit to the cave-house. This type of tourism, in the form of the cultural tour group, requires the preservation of cultural forms in some static state so that tourists can continue to be fed the stereotypes they are supposed to have come to see. By attempting to shroud the encounter in authenticity, the guide was achieving precisely the opposite.

In Göreme village, by contrast, the tourists can cope even with the presence of the highly 'modern' Internet, as long as they and the villagers are in a situation both to write their interactions and experiences partly for themselves and also to have interactions and experiences that are subject to happenstance. Tourists can cope with the paradoxes in their quests as long as they are free to negotiate them and to play with them; indeed, it is playing with the paradoxes and tourist identities that is part of the tourist's fun. For the independent tourists in Göreme village, regardless of whether the presentations and performances of the village and the villagers are perceived as authentically traditional, the encounters in themselves meet with the tourists' desires for the 'authentically social', precisely because they are not evidently staged. Selwyn's suggestion, then, that 'within the same individual tourist may beat a heart which is equally pilgrim-like and child-like' (Selwyn 1996: 6) seems to hold true, because the second gaze collapses the supposed dichotomies between authentic and contrived tourist experiences. In Göreme there is no longer an either/or situation, but rather it seems one where authenticity, on the one hand, and fantasy and fun, on the other, are inextricably mixed.

This point, then, goes some way towards unlocking one of the central paradoxes of cultural tourism, which is the truism that tourism necessarily destroys the object of its desire. The problem, as it was presented in the introduction chapter, is that discourses on culture and tourism tend to be characterised by a series of contradictions between tradition and tourism, and these contradictions are blocking the way forward for any notions of sustainability in cultural tourism. Rather than being the key to sustainable tourism, the aesthetic ideals that are carried under the guise of cultural sensitivity and asserted within preservation rhetoric might in fact be at the root of the problems and contradictions. If let to run, in other words, sustainable tourism ideology might itself threaten the sustainability of tourism in Göreme. We are reminded again here of Chambers' golf-course analogy and, further to this, Chambers argued:

> Resistance to change is as much an act of deliberateness as is the will to adopt new customs and practices. Authentic cultures might not be able to predict their futures or to act in a wholly independent manner, but they have the wherewithal to play a significant role in the participating in those processes that will shape their lives.

(Chambers 2000: 99)

Rather than constantly reiterating the need for preservation and protection of 'tradition' in some *a priori* form, there is a need to develop an understanding of where cultural tourism might begin to encompass and even embrace the new cultural forms that inevitably emerge through tourism itself. While new forms of cultural identity, such as the use of Flintstones imagery in Göreme's tourism, may appear to be a hybridisation of cultures and thus appear as 'Westernisation' or homogenisation, it is important to remember that the global is always negotiated through and by the local, and so these hybrid cultures always create a vernacular point of difference.

The conceptualisation of the 'sustainability', or the successful continuation, of cultural tourism needs to accommodate local social and environmental change, and recognise new cultural forms that emerge, rather than being entirely focused on preservation of some imagined stability. The Flintstones characterisation performed by the villagers permits a sense of equality in those villagers' interactions with tourists precisely because it openly acknowledges an awareness of representations of themselves and Göreme, and brings them together with tourists in a *communitas* of irony. It is also necessary to consider the extent to which tourists themselves can and will continue to be able to accommodate the inevitable change in the cultures and environments they visit. As we have seen, this play on Göreme cave-life works within a post-modern ethos to meet the tourist's desire to be in a place that is both real and yet fantastic at the same time, and to encounter people who are both 'authentically other' yet fun and fictional at the same time. Even the serious authenticity-seeking tourists can enjoy some fun, and this fun is plentiful in the fantasy of being troglodytes in a Flintstones-land – a fantasy that the tourists can play out together with their 'hosts' in the liminal zone of Göreme's tourism sphere.

In other words, tourists are probably far more pliable than we generally think. They are very much 'post-tourists' (Feifer 1985) in that they are fully aware that they are tourists and that they are not invisible observers in an hitherto untouched Turkish village. Indeed, contemporary tourists have developed something of an ambivalence towards 'authenticity', certainly in the sense of authenticity as a quality embedded in the places and people they visit. Instead, they seek unplanned and unexpected encounters and experiences with those places and people. The authentic versus inauthentic dichotomy thus becomes less relevant when we consider the second tourist gaze, and the most important poles in typologies or continua regarding contemporary tourist experience might rather be serendipity, or unexpected experiences, versus those experiences that are pre-planned and predictable.

A pivotal issue here, then, is the varying levels and ways that tourists tap into the 'tourism industry', because certain aspects of the industry, and particularly the cultural guided tour, tend to structure tourist experiences and fix them on the primary tourist gaze. While tourists who are travelling independently of packaged tours are often overlooked by governments and tourism industry bodies because the income they bring is usually not taken as seriously as the income derived from mass group tourists (see Loker-Murphy and Pearce 1995), it has been shown in this book precisely why this particular type of cultural tourism should be taken more seriously. Not only might small-scale and locally owned tourism business bring more direct financial benefit to local communities, but it inspires relations between the tourists and the peoples and places they visit that tend towards a more dynamic notion of cultural sustainability, and consequently to a more beneficial tourism.

Living with tourism in Göreme: a postscript

Returning to Göreme for short visits in the summers of 2000 and 2001, the village seems, more than anything, a 'tourist site'. The dominant feature of any tourist landscape is the tourists themselves, and in Göreme they are everywhere. Hordes of backpackers wander the streets from early morning until the early hours of the next day, when they drunkenly make their way back from the Flintstones bar to the Flintstones or some other cave-*pansiyon*. The landscape itself is also increasingly constructed according to tourist images and myths: panoramic viewpoints are signposted; a specific sunset-viewing location has been designated; and rather than discovering apparently long-lost churches deep in a Göreme valley, tourists are increasingly led to them by signposts erected on footpaths by the National Park authorities.

There are now as many as twenty-five tour agencies and a handful of new *pansiyons* in the village, some of which have been refurbished to be a little more upmarket than others. The municipality workers are busy constructing a large new *belediye* building that looks over the busy bus station, itself recently extended and the row of tour agencies and ticketing offices re-surfaced with local stone. New restaurants and shops are open, and the mayor's 'improvements' in the village centre, such as fancy street lamps and garden areas planted with flowers, are all making Göreme appear less and less like the quiet dusty village that it appeared to be even just a few years ago. Cybercafés have also sprung up everywhere, and tourists emailing friends and family back in New Zealand while drinking cappuccino have become a new normality among the caves. Now the number of tour buses thundering through every day is so great that Abbas said, as we sat together looking out on the main road from the steps of his agency, '*Istanbul gibi oliyor*' (it's becoming like Istanbul).

One of the most important events in recent years was the opening in 1998 of a new international airport near Nevşehir, approximately twenty kilometres from Göreme. This airport is slowly reaching full working capacity, and when it does so it will clearly have a profound effect on the future of tourism in the Cappadocia region and villages such as Göreme. In particular, the airport may bring the possibility of a level and type of tourism that would be good for Göreme to move towards. Previously, the nearest airport was at Kayseri, situated over a hundred kilometres away, and the main choices available to tourists were either an all-in coach tour or the opposite extreme of 'backpacking' and having to find their own way to Göreme by public transport. With the new airport close by, and also increasing opportunities for Göreme businesses to promote and sell their tourism services directly to potential tourists via the Internet, some tourists from the package tour groups might be creamed off and drawn in to Göreme village to sample the delights of the more vernacular services on offer there. Indeed, some members of the American group who came out disgruntled from their cave-house visit might wish to fly to the Cappadocia Airport and pay slightly more for accommodation than the usual backpackers, but they might at the same time desire to engage the 'second gaze' with vernacular-style services and encountering 'local' experiences rather than being shown cultural stereotypes on an all-in package tour.

This fits rather well with an apparent global trend in the purchase of tourism products away from the uniformity and predictability of the package tour and towards more specialised forms of tourism. Articles in the travel pages of the British press convey this trend:

> The package holiday will soon be gone, according to Britain's biggest tour operator...Out goes the traditional welcome meeting, sugary cocktails and usual excursions. In comes 'a range of different welcomes' and 'interactive Millennium experiences'.
>
> (*Daily Telegraph*, 24 July 1999)

Such tourist experiences are certainly available in Göreme within the 'cavey' vernacular architecture, and from interactions with the local place and people that suit tourists' desires for more individualised experiences. Meanwhile, the Göreme villagers would be able to continue to prosper financially and to continue offering hospitality to their tourist guests in such a way as to retain a sense of place and control in their village. Overall, then, the future for Göreme is looking pretty good, and a sense of hope is especially strong among the younger entrepreneurs today. In contrast to the older entrepreneurs, who are beginning to express a nostalgia for the earlier days when they seemed as individuals to have more control in the tourism processes, I heard many of the young men planning what they will do and how tourism in Göreme will be when they take charge and become mayor.

On a more cautionary note, however, while I have argued throughout this book why the 'independent' type of tourism is more successful and conducive to cultural sustainability in that it allows for local ownership and control of tourism business, and thus for negotiations of identity and relationships, it should be remembered that *Living with Tourism* and the rapidity of social change it evokes is far from easy. The troubles and strife that occur in the context of tourism are sometimes so intense in Göreme that the balance between resilience and sense of control, on the one hand, and fragility and downright confusion, on the other, can seem extremely rocky.

It might also be remembered here just how fickle the tourism industry can be. During the summer of 1999, the situation in Göreme seemed fairly close to desperation as the war in Kosovo and the troubles surrounding the Kurdish problem were enough to keep most tourists away. To finish off that year came the devastating earthquake in the north-west part of Turkey. Tourism was just beginning to pick up again in the summers of 2000 and 2001, only to be struck another blow in that latter year by the events of September 11. While the effect of those events was reportedly not as bad in Göreme as might have been expected, once again the tugs and tensions between the global and the local have clearly been impressed on the villagers who are attempting to make their living from tourism. Perhaps, then, both the tourists *and* the local people need to develop a liking for the serendipity in their circumstances, because it seems clear that much of the complexity of contemporary tourism comes down to chance and thus requires a great deal of sagacity from all involved.

Bibliography

AA (1994) *Essential Guide to Turkey*, Basingstoke: AA Publishing.

—— (1995) *Essential Explorer Guide to Turkey*, Basingstoke: AA Publishing.

Abadan-Unat, N. (1986) *Women in the Developing World: Evidence from Turkey*, Colorado: University of Denver.

—— (1993) 'Impact of External Migration on Rural Turkey', in P. Stirling (ed.), *Culture and Economy – Changes in Turkish Villages*, Huntingdon: Eothen Press.

Abou-Zeid, A.M. (1965) 'Honour and Shame Among the Bedouins of Egypt', in J.G. Peristiany (ed.), *Honour and Shame – The Values of Mediterranean Society*, London: Weidenfeld & Nicolson.

Abram, S. (1996) 'Reactions to Tourism: A View from the Deep Green Heart of France', in J. Boissevain (ed.), *Coping with Tourists*, Oxford: Berghahn Books.

—— (1997) 'Performing for Tourists in Rural France', in S. Abram, D. Macleod and J. Waldren (eds), *Tourists and Tourism – identifying with people and places*, Oxford: Berg.

Abram, S., Macleod, D. and Waldren, J. (eds) (1997) *Tourists and Tourism – identifying with people and places*, Oxford: Berg.

Adams, K.M. (1984) 'Come to Tana Toraja, "Land of the Heavenly Kings": Travel Agents as Brokers in Ethnicity', *Annals of Tourism Research* 11: 469–85.

—— (1990) 'Cultural Commoditization in Tana Toraja, Indonesia', *Cultural Survival Quarterly* 14: 31–3.

Adler, J. (1989) 'Origins of Sightseeing', *Annals of Tourism Research* 16: 7–29.

Adventures Great and Small (2002) 'Unusual Places/Cultures: Cavetowns and Gorges of Cappadocia', *http://www.great-adventures.com/destinations/turkey/cappadocia.html* (accessed 15 March 2002).

Ahmad, F. (1993) *The Making of Modern Turkey*, London: Routledge.

Ahmed, A. and Shore, C. (1995) 'Introduction: Is Anthropology Relevant to the Contemporary World?', in A. Ahmed and C. Shore (eds), *The Future of Anthropology – Its Relevance to the Contemporary World*, London: The Athlone Press.

Allcock, J. (1995) 'International Tourism and the Appropriation of History in the Balkans', in M. Lanfant, J. Allcock and E. Bruner (eds), *International Tourism – Identity and Change*, London: Sage.

Ames, M. (1992) *Cannibal Tours and Glass Boxes – The Anthropology of Museums*, Vancouver: UBC Press.

Amfora (1995) Cappadocia tourism promotional magazine, Ankara: Ajansmat Matbaacilik.

Anderson, B. (1983) *Imagined Communities*, London: Verso.

Ardener, S. (1989) '"Remote Areas" – Some Theoretical Considerations', in N. Chapman (ed.), *Edwin Ardener: The Voice of Prophecy and Other Essays*, Oxford: Basil Blackwell.

Ari, O. (ed.) (1977) *Readings in Rural Sociology*, Istanbul: Bogaziçi University.

Arnold, G. (1989) *Journey Round Turkey*, London: Cassell.

Bailey, F.G. (1969) *Stratagems and Spoils – The Social Anthropology of Politics*, Oxford: Basil Blackwell.

—— (ed.) (1971) *Gifts and Poison – The Politics of Reputation*, Oxford: Basil Blackwell.

Balakrishnan, G. (1996) 'The National Imagination', in G. Balakrishnan (ed.), *Mapping the Nation*, London: Verso.

Baudrillard, J. (1983) *Simulations*, New York: Semiotext(e).

Bauman, R. (ed.) (1992) *Folklore, Cultural Performances and Popular Entertainments*, Oxford: Oxford University Press.

Bauman, Z. (1996) 'From Pilgrim to Tourist – or a Short History of Identity', in S. Hall and P. du Gay (eds), *Questions of Cultural Identity*, London: Sage.

Bazin, C. (1995) 'Industrial Heritage in the Tourism Process in France', in M. Lanfant, J. Allcock and E. Bruner (eds), *International Tourism – Identity and Change*, London: Sage.

Bedding, J. (1999) 'Turkey '99', *Telegraph Travel*, 30 January.

Bell, D. (1993) 'Yes Virginia, there is a feminist ethnography: reflections from three Australian fields', in D. Bell, P. Caplan and W.J. Karim (eds), *Gendered fields: women, men and ethnography*, London: Routledge.

Bellér-Hann, I. and Hann, C. (2001) *Turkish Region: State, Market and Social Identities on the East Black Sea Coast*, Oxford: James Curry.

Berger, J. (1972) *Ways of Seeing*, London: British Broadcasting Corporation and Penguin Books.

Berger, P. (1974) *The Homeless Mind*, Harmondsworth: Penguin.

Bernard, H.R. (1994) *Research Methods in Anthropology*, 2nd edn, London: Sage.

Berno, T. (1999) 'When a Guest is a Guest: Cook Islanders View Tourism', *Annals of Tourism Research* 26: 656–75.

Bezmen, C. (1996) *Tourism and Islam in Cappadocia*, unpublished Ph.D. thesis, Cambridge University, Cambridge.

Black, A. (1996) 'Negotiating the Tourist Gaze: The Example of Malta', in J. Boissevain (ed.), *Coping with Tourists*, Oxford: Berghahn Books.

Blair, J. (1970) 'Keeping House in a Cappadocian Cave', *National Geographic*, July: 127–46.

Blok, A. (1981) 'Rams and Billy-Goats: A Key to the Mediterranean Code of Honour', *Man* 16: 427–40.

Blue Guide to Turkey (1995) London: A & C Black.

Boissevain, J. (1979) 'Towards a Social Anthropology of the Mediterranean', *Current Anthropology* 20: 81–5.

—— (1996a) 'Ritual, Tourism and Cultural Commoditization in Malta: Culture by the Pound?', in T. Selwyn (ed.), *The Tourist Image: Myths and Myth-Making in Tourism*, New York and London: John Wiley & Sons.

—— (ed.) (1996b) *Coping with Tourists*, Oxford: Berghahn Books.

Boniface, P. (2001) *Dynamic Tourism: Journeying with Change*, Clevedon: Channel View Publications.

Boorstin, D. (1961) *The Image: A Guide to Pseudo-Events in America*, New York: Harper & Row.

Bourassa, S.C. (1991) *The Aesthetics of Landscape*, London and New York: Belhaven Press.

Bourdieu, P. (1966) 'The Sentiment of Honour in Kabyle Society', in J.G. Peristiany (ed.), *Honour and Shame – The Values of Mediterranean Society*, London: Weidenfeld & Nicolson.

—— (1977) *Outline of a Theory of Practice*, Cambridge: Cambridge University Press.

—— (1984) *Distinction*, London: Routledge.

—— (1990) *In Other Words*, Oxford: Polity Press.

Bowman, G. (1989) 'Fucking Tourists: Sexual Relations and Tourists in Jerusalem's Old City', *Critical Anthropology* IX: 77–93.

—— (1992) 'The Politics of Tour Guiding: Israeli and Palestinian guides in Israel and the Occupied Territories', in D. Harrison (ed.), *Tourism and the Less Developed Countries*, London: Belhaven.

—— (1996) 'Passion, Power and Politics in a Palestinian Tourist Market', in T. Selwyn (ed.), *The Tourist Image: Myths and Myth-Making in Tourism*, New York and London: John Wiley & Sons.

Bramwell, B. and Lane, B. (eds) (1994) *Tourism and Sustainable Rural Development*, Clevedon: Channel View Publications.

Breger, R. and Hill, R. (eds) (1998) *Cross-Cultural Marriage*, Oxford: Berg.

Brewer, J. (2000) *Ethnography*, Buckingham: Open University Press.

Brook, S. (1996) 'It's a tufa life inside a rock dwelling', *The Times*, 24 February.

Brown, D. (1996) 'Genuine Fakes', in T. Selwyn (ed.), *The Tourist Image: Myths and Myth-Making in Tourism*, New York and London: John Wiley & Sons.

Brown, M.F. (1998) 'Can Culture Be Copyrighted?', *Current Anthropology* 39: 193–222.

Bruner, E. (1989) 'Of Cannibals, Tourists and Ethnographers', *Cultural Anthropology* 4: 438–45.

—— (1991) 'Transformation of Self in Tourism', *Annals of Tourism Research* 18: 238–50.

—— (1995) 'The Ethnographer/Tourist in Indonesia', in M. Lanfant, J. Allcock and E. Bruner (eds), *International Tourism – Identity and Change*, London: Sage.

—— (1996a) 'Experience and its Expressions', in V. Turner and E. Bruner (eds), *The Anthropology of Experience*, Chicago: University of Illinois Press.

—— (1996b) 'Tourism in the Balinese Borderzone', in S. Lavie and T. Swedenburg (eds), *Displacement, Diaspora and Geographies of Identity*, Durham: Duke University Press.

Burke, M.M. (1998) 'The "Dawn" of Cappadocia – a wild place to be!', *Turkish Daily News*, 31 March.

Butler, K. (1995) 'Independence for Western Women through Tourism', *Annals of Tourism Research* 22: 487–9.

Butler, R. and Hinch, T. (eds) (1996) *Tourism and Indigenous Peoples*, London: International Thomson Business Press.

Butler, R. and Pearce, D. (eds) (1995) *Change in Tourism – People, Places, Processes*, London: Routledge.

Butler, R. and Stiakaki, E. (2001) 'Tourism and Sustainability in the Mediterranean: Issues and Implications from Hydra', in D. Ioannides, Y. Apostolopoulos and S. Sonmez, *Mediterranean Islands and Sustainable Tourism Development: Practices, Management and Policies*, London: Continuum.

Buzard, J. (1993) *The Beaten Track: European Tourism, Literature and the Ways to Culture*, Oxford: Clarendon Press.

Campbell, C. (1987) *The Romantic Ethic and the Spirit of Modern Consumerism*, Oxford: Basil Blackwell.

Campbell, J. (1964) *Honour, Family and Patronage*, New York: Oxford University Press.

Cappadocia (1995) Guidebook (English edn), Istanbul: NET.

Carrier, J. (1995a) 'Introduction', in J. Carrier (ed.), *Occidentalism – Images of the West*, Oxford: Clarendon Press.

—— (1995b) 'Maussian Occidentalism: Gift and Commodity Systems', in J. Carrier (ed.), *Occidentalism – Images of the West*, Oxford: Clarendon Press.

Carrithers, M., Collins, S., and Lukes, S. (eds) (1987) *The Category of the Person. Anthropology, Philosophy, History*, Cambridge: Cambridge University Press.

Carter, E., Donald, J. and Squires, J. (1993) *Space and Place – Theories of Identity and Location*, London: Lawrence & Wishart.

Castelberg-Koulma, M. (1991) 'Greek Women and Tourism: Women's Agro-Tourist Co-operatives as an Alternative Form of Organisation', in N. Redcliff and M.T. Sinclair (eds), *Working Women: International Perspectives on Labour and Gender Ideology*, London: Routledge.

Chambers, E. (ed.) (1997) *Tourism and Culture – An Applied Perspective*, New York: State University of New York Press.

—— (2000) *Native Tours: The Anthropology of Travel and Tourism*, Prospect Heights: Waveland Press.

Chambers, R.E. and McBeth, M.K. (1992) 'Community encouragement: Returning to the basis for community development', *Journal of the Community Development Society* 23: 20–38.

Chapman, N. (ed.) (1989) *Edwin Ardener: The Voice of Prophecy and Other Essays*, Oxford: Basil Blackwell.

Cincone, L. (1988) *The Role of Development in the Exploitation of Southeast Asian Women: Sex Tourism in Thailand*, New York: Women's International Resource Exchange.

Classen, C. (1993) *Worlds of Sense*, London: Routledge.

Classen, C., Howes, D. and Synnott, A. (1994) *Aroma: the cultural history of smell*, London: Routledge.

Clifford, J. (1997) *Routes: Travel and Translation in the Late Twentieth Century*, Cambridge, MA: Harvard University Press.

—— (1998) *The Predicament of Culture: Twentieth-Century Ethnography, Literature and Art*, Cambridge, MA: Harvard University Press.

Cohen, A. (1985) *The Symbolic Construction of Community*, London: Routledge.

Cohen, E. (1971) 'Arab Boys and Tourist Girls in a Mixed Jewish-Arab Community', *International Journal of Comparative Sociology* XII: 217–33.

—— (1979) 'A Phenomenology of Tourist Experiences', *Sociology* 13: 179–201.

—— (1982) 'Marginal Paradises: Bungalow Tourism on the Islands of Southern Thailand', *Annals of Tourism Research* 9: 189–228.

—— (1988a) 'Traditions in the Qualitative Sociology of Tourism', *Annals of Tourism Research* 15: 29–46.

—— (1988b) 'Authenticity and Commoditization in Tourism', *Annals of Tourism Research* 15: 371–86.

—— (1989) ' "Primitive and Remote": Hill Tribe Trekking in Thailand', *Annals of Tourism Research* 16: 30–61.

—— (1995) 'Contemporary Tourism – trends and challenges: sustainable authenticity or contrived post-modernity?', in R. Butler and D. Pearce (eds), *Change in Tourism – People, Places, Processes*, London: Routledge.

Cohen, S. and Taylor, L. (1992) *Escape Attempts: The Theory and Practice of Resistance to Everyday Life*, 2nd edn, London: Routledge.

Coleman, S. and Elsner, J. (1998) 'Performing Pilgrimage: Walsingham and the Ritual Construction of Irony', in F. Hughes-Freeland (ed.), *Ritual, Performance, Media*, London: Routledge.

Cone, C.A. (1995) 'Crafting Selves – The Lives of Two Mayan Women', *Annals of Tourism Research* 22: 314–27.

Craik, J. (1995) 'Are There Cultural Limits to Tourism?', *The Journal of Sustainable Tourism* 3: 87–98.

Crawshaw, C. and Urry, J. (1997) 'Tourism and the Photographic Eye', in C. Rojek and J. Urry (eds), *Touring Cultures: Transformations of Travel and Theory*, London: Routledge.

Crick, M. (1985) 'Tracing the Anthropological Self: Quizzical Reflections on Fieldwork, Tourism and the Ludic', *Social Analysis* 17: 71–92.

—— (1989) 'Representations of sun, sex, sights, savings and servility: international tourism in the social sciences', *Annual Review of Anthropology* 18: 307–44.

—— (1992) 'Ali and me: an essay in street-corner anthropology', in J. Okely and H. Callaway (eds), *Anthropology and Autobiography*, London: Routledge.

—— (1994) *Resplendent Sites, Discordant Voices: Sri Lankans and international tourism*, Chur, Switzerland: Harwood Academic Publishers.

—— (1995) 'The Anthropologist as Tourist: an Identity in Question', in M. Lanfant, J. Allcock and E. Bruner (eds), *International Tourism – Identity and Change*, London: Sage.

Crouch, D. (1994) 'Home, Escape and Identity: Rural Cultures and Sustainable Tourism', in B. Bramwell and B. Lane (eds), *Rural Tourism and Sustainable Rural Development*, Clevedon: Channel View Publications.

Csordas, T. (1994) *Embodiment and Experience*, Cambridge: Cambridge University Press.

Daily Telegraph (1999) 24 July.

Dann, G. (1996) *The Language of Tourism – A Sociolinguistic Perspective*, Wallingford: CAB International.

—— (1999) 'The People of Tourist Brochures', in T. Selwyn (ed.), *The Tourist Image: Myths and Myth-Making in Tourism*, New York and London: John Wiley & Sons, 1996.

Davies, C.A. (1999) *Reflexive Ethnography*, London: Routledge.

De Kadt, E. (ed.) (1979) *Tourism – Passport to Development?*, Washington, DC: Oxford University Press (published for the World Bank and UNESCO).

Delaney, C. (1991) *The Seed and the Soil – Gender and Cosmology in Turkish Village Society*, Berkeley and Los Angeles: University of California Press.

—— (1993) 'Traditional Modes of Authority and Co-operation', in P. Stirling (ed.), *Culture and Economy – Changes in Turkish Villages*, Huntingdon: Eothen Press.

Drost, A. (1996) 'Developing Sustainable Tourism for World Heritage Sites', *Annals of Tourism Research* 23: 479–92.

Droste, B. (1995) 'Cultural Landscapes in a Global World Heritage Strategy', in B. Droste, H. Plachter and M. Rossler (eds), *Cultural Landscapes of Universal Value*, Jena: Gustav Fischer.

Echtner, C. (1995) 'Entrepreneurial Training in Developing Countries', *Annals of Tourism Research* 22: 119–34.

Eco, U. (1986) *Travels in Hyperreality*, Orlando: Harcourt Brace & Co.

Elias, N. and Dunning, E. (1986) *Quest for Excitement*, Oxford: Basil Blackwell.

Elias, N. and Scotson, J. (1994) *The Established and the Outsiders*, London: Sage.

Ellen, R.F. (ed.) (1984) *Ethnographic Research: A Guide to General Conduct*, London: Academic Press.

Elsrud, T. (2001) 'Risk Creation in Travelling: Backpacker Adventure Narration', *Annals of Tourism Research* 28(3): 597–617.

Emge, A. (1990) *Wohnen in den Hohlen von Göreme – Traditionelle Bauweise und Symbolik in Zentralanatolien*, Ph.D. thesis, Berlin: Dietrich Reimer Verlag.

—— (1992) 'Old Order in New Space: Change in the Troglodytes' Life in Cappadocia', in *Change in Traditional Habitat*, Traditional Dwellings and Settlements Working Paper Series, vol. 37, Berkeley: University of California.

Engelbrektsson, U.B. (1978) *The Force of Tradition – Turkish Migrants at Home and Abroad*, Sweden: Acta Universitatis Gothoburgensis.

Erdentug, N. (1977) 'Family Structure and Marriage Customs of a Turkish Village', in O. Ari (ed.), *Readings in Rural Sociology*, Istanbul: Bogaziçi University.

Erdogdu, D. (2002) 'Online Cappadocia', *http://www.captr.com/cappadocia.htm* (accessed 27 March 2002).

Errington, F. and Gewertz, D. (1989) 'Tourism and Anthropology in a Post-modern World', *Oceana* 60: 37–54.

Explore Worldwide (1989) *Turkey* (promotional video).

Fagence, M. (1998) 'Rural and Village Tourism in Developing Countries', *Third World Planning Review* 20: 107–18.

Falk, P. (1994) *The Consuming Body*, London: Sage.

Fardon, R. (ed.) (1990) *Localising Strategies – Regional Traditions of Ethnographic Writing*, Edinburgh: Scottish Academic Press.

Faulkenberry, L., Coggeshall, J., Backman, K. and Backman, S. (2000) 'A Culture of Servitude: The Impact of Tourism and Development on South Carolina's Coast', *Human Organization* 59: 86–95.

Featherstone, M. (1991) *Consumer Culture and Postmodernism*, London: Sage.

Feifer, N. (1985) *Going Places*, London: Macmillan.

Fetterman, D. (1998) *Ethnography*, London: Sage.

Fiertz, G. (1996) 'Tourism Development in Turkey', in A. Badger *et al.* (eds), *Trading Places: Tourism as Trade*, London: Tourism Concern.

Fine, G.A. (1995) 'Public Narrations and Group Culture: Discerning Discourse in Social Movements', in H. Johnston and B. Klandermans (eds), *Social Movements and Culture*, London: UCL Press.

Finkel, A. and Sirman, N. (eds) (1990) *Turkish State, Turkish Society*, London and New York: Routledge.

Foster, G. (1965) 'Peasant Society and the Image of Limited Good', *American Anthropologist* 67: 293–315.

Foucault, M. (1977) *Discipline and Punish: the birth of the prison*, London: Allen Lane.

Fruzzetti, L.M. (1982) *The Gift of a Virgin*, Delhi: Oxford University Press.

Game, A. (1991) *Undoing the Social. Towards a Deconstructive Sociology*, Buckingham: Open University Press.

Garcia-Ramon, M.D., Canoves, G. and Valdovinos, N. (1995) 'Farm Tourism, Gender and the Environment in Spain', *Annals of Tourism Research* 22: 267–92.

Gilmore, D. (ed.) (1987) *Honour and Shame and the Unity of the Mediterranean*, Washington, DC: American Anthropological Association.

Gilsenan, M. (1990) 'Very Like a Camel: The Appearance of an Anthropologist's Middle East', in R. Fardon (ed.), *Localising Strategies – Regional Traditions of Ethnographic Writing*, Edinburgh: Scottish Academic Press.

—— (1992) *Lords of the Lebanese Marches – Violence and Narrative in Arab Society*, London: I.B. Tauris.

Goddard, V. (1994) 'From the Mediterranean to Europe: Honour, Kinship and Gender', in V. Goddard, J. Llobera and C. Shore (eds), *The Anthropology of Europe*, Oxford: Berg.

Goddard, V., Llobera, J. and Shore, C. (eds) (1994) *The Anthropology of Europe*, Oxford: Berg.

Goffman, E. (1959) *Presentation of Self in Everyday Life*, New York: Doubleday.

—— (1967) *Interaction Ritual*, London: Allen Lane/The Penguin Press.

Golde, P. (ed.) (1986) *Women in the Field – Anthropological Perspectives*, London: University of California Press.

Goody, J. (1977) *The Domestication of the Savage Mind*, Cambridge: Cambridge University Press.

Göreme National Park Group (late 1960s) *Göreme National Park Plan*, Ankara.

Graburn, N. (1983) 'The Anthropology of Tourism', *Annals of Tourism Research* 10: 9–31.

—— (1989) 'Tourism: The Sacred Journey', in V. Smith (ed.), *Hosts and Guests*, Philadelphia: University of Pennsylvania Press.

Green, N. (1990) *The Spectacle of Nature. Landscape and bourgeois culture in nineteenth-century France*, Manchester and New York: Manchester University Press.

Greenfield, S., Strickon, A. and Aubey, R. (eds) (1979) *Entrepreneurs in Cultural Context*, Albuquerque: University of New Mexico Press.

Greenwood, D.J. (1989) 'Culture by the Pound: An Anthropological Perspective on Tourism as Cultural Commoditization', in V. Smith (ed.), *Hosts and Guests*, Philadelphia: University of Pennsylvania Press.

Gregory, D. (1999) 'Scripting Egypt: Orientalism and the Cultures of Travel', in J. Duncan and D. Gregory (eds), *Writes of Passage: Reading Travel Writing*, London and New York: Routledge.

Grewal, I. (1996) *Home and Harem: nation, gender, empire and the cultures of travel*, London: Leicester University Press.

Gronow, J. (1997) *The Sociology of Taste*, London: Routledge.

Haber, W. (1995) 'Concept, Origin and Meaning of "Landscape"', in B. Droste, H. Plachter and M. Rossler (eds), *Cultural Landscapes of Universal Value*, Jena: Gustav Fischer.

Hall, C.M. (1992) 'Sex tourism in South-east Asia', in D. Harrison (ed.), *Tourism and the Less Developed Countries*, London: Belhaven.

Hall, C.M. and Lew, A. (eds) (1998) *Sustainable Tourism – A Geographical Perspective*, Harlow: Longman.

Hall, C.M. and McArthur, S. (eds) (1993) *Heritage Management in New Zealand and Australia*, Oxford: Oxford University Press.

Hall, S. and du Gay, P. (eds) (1996) *Questions of Cultural Identity*, London: Sage.

Hampton, M. (1998) 'Backpacker Tourism and Economic Development', *Annals of Tourism Research* 25: 639–60.

Harkin, M. (1995) 'Modernist Anthropology and Tourism of the Authentic', *Annals of Tourism Research*, 22(3): 650–70.

Harrison, D. (1992) *Tourism and the Less Developed Countries*, London: Belhaven.

Harrison, S. (1999) 'Cultural Boundaries', *Anthropology Today* 15: 10–13.

Hart, K. (1973) 'Informal Income Opportunities and Urban Employment in Ghana', *Journal of Modern African Studies* 11: 61–89.

Hastrup, K. (1987) 'Fieldwork among friends: ethnographic exchange within the northern civilisation', in A. Jackson (ed.), *Anthropology at Home*, London: Tavistock.

—— (1992) 'Writing Ethnography', in J. Okely and H. Callaway (eds), *Anthropology and Autobiography*, London: Routledge.

Hastrup, K. and Hervik, P. (eds) (1994) *Social Experience and Anthropological Knowledge*, London: Routledge.

Heal, F. (1990) *Hospitality in Early Modern England*, Oxford: Clarendon Press.

Helms, M. (1988) *Ulysses' Sail*, Princeton: Princeton University Press.

Herold, E., Garcia, R. and DeMoya, T. (2001) 'Female Tourists and Beach Boys: Romance or Sex Tourism?', *Annals of Tourism Research* 28(4): 978–97.

Hertz, R. (1997) *Reflexivity and Voice*, London: Sage.

Herzfeld, M. (1980) 'Honour and Shame: Problems in the Comparative Analysis of Moral Systems', *Man* 15: 339–51.

—— (1987) ' "As In Your Own House": Hospitality, Ethnography, and the Stereotype of Mediterranean Society', in D. Gilmore (ed.), *Honour and Shame and the Unity of the Mediterranean*, Washington, DC: American Anthropological Association.

—— (1991) *Place in History: Social and Monumental Time in a Cretan Town*, Princeton: Princeton University Press.

Higuchi, T. (1989) *The Visual and Spatial Structure of Landscapes*, Cambridge, MA, and London: MIT Press.

Hirsch, E. and O'Hanlon, M. (eds) (1995) *The Anthropology of Landscape – perspectives on place and space*, Oxford: Clarendon Press.

Hirschon, R. (ed.) (1983) *Women and Property – Women as Property*, New York: St Martin's Press.

Hitchcock, M., King, V.T. and Parnwell, M. (eds) (1993) *Tourism in South-East Asia*, London: Routledge.

Hodson, M. (1999) 'Hidden Turkey', *The Sunday Times*, 9 May.

Hollinshead, K. (1997) 'Heritage Tourism under Postmodernity: Truth and the Past', in C. Ryan (ed.), *Tourist Experience*, London: Cassell.

Holmes, H. (1996) 'Gone to Turkey Underground', in *Discovery Communication, Inc.*, USA: Discovery Channel Online, www.discovery.com.

—— (1997) 'Falling Through the Cracks', *Escape* (US travel magazine): 55–61.

Howes, D. (1991) *The Varietes of Sensory Experience*, Toronto: University of Toronto Press.

Hufford, M. (ed.) (1994) *Conserving Culture – A New Discourse on Heritage*, Urbana and Chicago: University of Illinois Press.

Hughes, G. (1995) 'Authenticity in Tourism', *Annals of Tourism Research* 22: 781–803.

Ingram, P.T. (1990) *Indigenous Entrepreneurship and Tourism Development in the Cook Islands and Fiji*, unpublished Ph.D. thesis, Massey University, New Zealand.

Ioannides, D., Apostolopoulos, Y. and Sonmez, S. (2001) *Mediterranean Islands and Sustainable Tourism Development: Practices, Management and Policies*, London: Continuum.

Jackson, A. (ed.) (1987) *Anthropology at Home*, London: Tavistock.

Jackson, J.E. (1995) ' "Déjà Entendu": The Liminal Qualities of Anthropological Field-notes', in J.V. Maanen (ed.), *Representation in Ethnography*, California: Sage.

Jafari, J. (1987) 'Tourism Models: The Sociocultural Aspects', *Tourism Management*, 8(2): 151–9.

Jamal, T. and Getz, D. (1995) 'Collaboration Theory and Community Tourism Planning', *Annals of Tourism Research* 22: 186–204.

Jameson, F. (1991) *Postmodernism – or the cultural logic of late capitalism*, Durham: Duke University Press.

Jenkins, R. (1992) *Pierre Bourdieu*, London and New York: Routledge.

Johnson, P. and Thomas, B. (1992) *Perspectives on Tourism Policy*, London and New York: Mansell.

Kabbani, R. (1986) *Europe's Myths of Orient*, London: Pandora.

Kandiyoti, D. (1985) *Women in Rural Production Systems: problems and policies*, Paris: UNESCO.

——(1991) *Women, Islam and the State*, London: Macmillan.

—— (ed.) (1996) *Gendering the Middle East*, New York: Syracuse University Press.

Kearney, M. (1996) *Reconceptualising the Peasantry*, Colorado and Oxford: Westview Press.

Keyder, C. (1993) 'The Genesis of Petty Commodity Production in Agriculture – the case of Turkey', in P. Stirling (ed.), *Culture and Economy – Changes in Turkish Villages*, Huntingdon: Eothen Press.

King, A. (ed.) (1991) *Contemporary Conditions for the Representation of Identity*, Basingstoke and London: Macmillan Education.

King, V. (ed.) (1992) *Tourism in Borneo – Papers from the Second Biennial International Conference, Kota Kinabalu*, Malaysia: Borneo Research Council Proceedings Series.

Kinnaird, V. and Hall, D. (eds) (1994) *Tourism: A Gender Analysis*, Chichester: Wiley.

Kinnaird, V., Kothari, U. and Hall, D. (1994) 'Tourism: Gender Perspectives', in V. Kinnaird and D. Hall (eds), *Tourism: A Gender Analysis*, Chichester: Wiley.

Kinzer, S. (1997) 'Live Better, Live Longer, in My Home, Sweet Cave!', *The New York Times International*, 2 May.

Kirshenblatt-Gimblett, B. (1998) *Destination Culture: Tourism, Museums, and Heritage*, Berkeley and Los Angeles: University of California Press.

Kirshenblatt-Gimblett, B. and Bruner, E. (1992) 'Tourism', in R. Bauman (ed.), *Folklore, Cultural Performances and Popular Entertainments*, Oxford: Oxford University Press.

Kocturk, T. (1992) *A Matter of Honour – Experiences of Turkish Women Immigrants*, London: Zed Books.

Kohn, T. (1994) 'Incomers and fieldworkers: a comparative study of social experience', in K. Hastrup and P. Hervik (eds), *Social Experience and Anthropological Knowledge*, London: Routledge.

—— (1997) 'Island Involvement and the Evolving Tourist', in S. Abram, D. Macleod and J. Waldren (eds), *Tourists and Tourism – identifying with people and places*, Oxford: Berg.

—— (1998) 'The Seduction of the Exotic: Notes on Mixed Marriage in East Nepal', in R. Breger and R. Hill (eds), *Cross-Cultural Marriage*, Oxford: Berg.

Kolars, J.F. (1977) 'Tradition, Season and Change in a Turkish Village', in O. Ari (ed.), *Readings in Rural Sociology*, Istanbul: Bogaziçi University.

Korzay, M. (1994) 'Turkish Tourism Development', in A.V. Seaton *et al.* (eds), *Tourism – The State of the Art*, Chichester: Wiley.

Korzay, M., Burcoğlu, N.K., Yarcan, S. and Unalan, D. (1998) *Proceedings of the International Conference on Heritage, Multicultural Attractions and Tourism*, vols I and II, Istanbul: Bogaziçi University.

Kottak, C.P. (1983) *Assault on Paradise: Social Change in a Brazilian Village*, New York: Random House.

Krippendorf, J. (1987) *The Holiday Makers. Understanding the impact of leisure and travel*, Oxford: Heinemann Professional Publishing.

Kruhse-Mount Burton, S. (1995) 'Sex Tourism and Traditional Australian Male Identity', in M. Lanfant, J. Allcock and E. Bruner (eds), *International Tourism – Identity and Change*, London: Sage.

Kulick, D. and Willson, M. (1995) *Taboo*, London: Routledge.

Lanfant, M. (1995) 'International Tourism, Internationalisation and the Challenge to Identity', in M. Lanfant, J. Allcock and E. Bruner (eds), *International Tourism – Identity and Change*, London: Sage.

Lanfant, M., Allcock, J. and Bruner, E. (eds) (1995) *International Tourism – Identity and Change*, London: Sage.

Larson, P., Freudenberger, M. and Wyckoff-Baird, B. (1997) *Lessons from the field: A review of World Wildlife Fund's experience with integrated conservation and development projects 1985–1996*, Washington, DC: World Wildlife Fund.

Lash, S. and Friedman, J. (eds) (1992) *Modernity and Identity*, Oxford: Basil Blackwell.

Lash, S. and Urry, J. (1994) *Economies of Signs and Space*, London: Sage.

Lea, J. (1988) *Tourism and Development in the Third World*, New York: Routledge.

Levy, M. (1967) 'Social Patterns (Structures) and Problems of Modernization', in W. Moore and R. Cook (eds), *Readings on Social Change*, Englewood Cliffs, NJ: Prentice-Hall.

Leyssen, N. and Idiz, M. (1993) 'Göreme – Saving a Unique Historical Combination', *Turkish Daily News*, 16 May: 12–15.

Liechty, M. (1995) 'Kathmandu as Translocality: Multiple Places in a Nepali Space', in P. Yaeger (ed.), *The Geography of Identity*, Ann Arbor: University of Michigan Press.

Little, K. (1991) 'On Safari: The Visual Politics of a Tourist Representation', in D. Howes (ed.), *The Varieties of Sensory Experience*, Toronto: University of Toronto Press.

Loizos, P. and Papataxiarchis, E. (eds) (1991) *Contested Identities*, Princeton: Princeton University Press.

Loker-Murphy, L. and Pearce, P. (1995) 'Young Budget Travellers: Backpackers in Australia', *Annals of Tourism Research* 22: 819–43.

Lonely Planet – Travel Survival Kit to Turkey (1990) Victoria and Berkeley: Lonely Planet Publications.

—— (1996) Victoria and Berkeley: Lonely Planet Publications.

Lowenthal, D. (1998) *The Heritage Crusade and the Spoils of History*, Cambridge: Cambridge University Press.

Lutz, C. and Collins, J. (1993) *Reading National Geographic*, Chicago and London: University of Chicago Press.

—— (1994) 'The Photograph as an Intersection of Gazes – The Example of the *National Geographic*', in L. Taylor (ed.), *Visualising Theory*, London: Routledge.

Maanen, J.V. (ed.) (1988) *Tales of the Field: On Writing Ethnography*, Chicago: University of Chicago Press.

—— (ed.) (1995) *Representation in Ethnography*, California: Sage.

MacCannell, D. (1976) *The Tourist*, London: Macmillan.

—— (1989) 'Introduction (to Semiotics)', *Annals of Tourism Research* 16: 1–6.

—— (1992) *Empty Meeting Grounds – The Tourist Papers*, London and New York: Routledge.

—— (1994) 'Cannibal Tours', in L. Taylor (ed.), *Visualising Theory*, London: Routledge.

—— (2001) 'Tourist Agency', *Tourist Studies* 1(1): 23–37.

Macdougall, D. (1994) 'Whose Story Is It?', in L. Taylor (ed.), *Visualising Theory*, London: Routledge.

Macleod, D. (1997) ' "Alternative" Tourists on a Canary Island', in S. Abram, D. Macleod and J. Waldren (eds), *Tourists and Tourism – identifying with people and places*, Oxford: Berg.

Magnarella, P.J. (1974) *Tradition and Change in a Turkish Town*, Cambridge, MA: Schenkman Publishing.

Manning, F. (1992) 'Spectacle', in R. Bauman (ed.), *Folklore, Cultural Performances and Popular Entertainments*, Oxford: Oxford University Press.

Mansberger, M. (1995) 'Tourism and Cultural Change in Small-Scale Societies', *Human Organization* 54: 87–94.

Mansur, F. (1972) *Bodrum: A Town in the Aegean*, Leiden: E.J. Brill.

Marcus, J. (1992) *A World of Difference – Islam and Gender Hierarchy in Turkey*, London: Zed Books.

Mauss, M. (1990) *The Gift*, London: Routledge.

McClintock, A., Mufti, A. and Shohat, E. (1997) *Dangerous Liaisons: gender, nations and postcolonial perspectives*, Minneapolis: University of Minnesota Press.

Meethan, K. (1996) 'Place, Image and Power: Brighton as a Resort', in T. Selwyn (ed.), *The Tourist Image – Myths and Myth-Making in Tourism*, Chichester: John Wiley & Sons.

Meisch, L. (1995) 'Gringas and Otavalenos – Changing Tourist Relations', *Annals of Tourism Research* 22: 441–62.

Mellinger, W.M. (1994) 'Toward a Critical Analysis of Tourism Representations', *Annals of Tourism Research* 21: 756–79.

Michaud, J. (1991) 'A Social Anthropology of Tourism in Ladakh, India', *Annals of Tourism Research* 18: 605–21.

Miller, D. (ed.) (1995) 'Introduction', *Worlds Apart – Modernity through the prism of the local*, London: Routledge.

Milli Parklar Dairesi Baskanligi (n.d.) *Göreme Historical National Park* – leaflet produced by the National Park Directorate.

Milton, K. (ed.) (1993) *Environmentalism – The View from Anthropology*, ASA Monograph 32, London: Routledge.

Ministry of Tourism of the Republic of Turkey (1994) *Cappadokya* (promotional magazine).

—— (1994) *Turkey* (promotional magazine).

Misztal, B. (1996) *Trust*, Cambridge: Polity Press.

Mızrak, M. (1995) 'Göreme: An Epic from Nature', *Amfora* (Cappadocia tourism promotion magazine), Ankara: Ajansmat Matbaacilik.

Moore, H.L. (1994) *A Passion for Difference*, Cambridge: Polity Press.

Moore, W. and Cook, R. (eds) (1967) *Readings on Social Change*, Englewood Cliffs: Prentice-Hall.

Morris, C. (1998) 'Troglodyte Turks seek refuge from urban blight', *Guardian*, 5 October.

Moscardo, G.M. and Pearce, P.L. (1986) 'Historic Theme Parks: an Australian Experience of Authenticity', *Annals of Tourism Research* 13(3): 467–79.

Mowforth, M. and Munt, I. (1998) *Tourism and Sustainability – New Tourism in the Third World*, London: Routledge.

Mumford, L. (1934) *Technics and Civilization*, New York: Harcourt, Brace & World.

Munt, I. (1994a) 'The "other" postmodern tourism: culture, travel and the new middle classes', *Theory, Culture and Society* 11: 101–23.

—— (1994b) 'Eco-tourism or ego-tourism?', *Race and Class* 36: 49–60.

Murphy, P. (1985) *Tourism: a community approach*, New York: Methuen.

Nash, D. (1995) 'An exploration of tourism as superstructure', in R. Butler and D. Pearce (eds), *Change in Tourism – People, Places, Processes*, London: Routledge.

—— (1996) *Anthropology of Tourism*, Oxford: Pergamon.

Nash, D. and Smith, V. (1991) 'Anthropology and Tourism', *Annals of Tourism Research* 18: 12–25.

Nation, R.C. (1996) 'Preface', in V. Mastny and R.C. Nation (eds), *Turkey Between East and West: New Challenges for a Rising Regional Power*, Colorado: Westview Press.

National Geographic (1939) 'Cones of Cappadocia', illustrations E. Matson, LXXVI (December): i–xvi.

—— (1958) 'Cappadocia: Turkey's Country of Cones', illustrations M. Riboud (January): 122–46.

Nunez, T.A. (1977) 'Touristic studies in anthropological perspectives', in V. Smith (ed.), *Hosts and Guests: An Anthropology of Tourism*, Philadelphia: University of Pennsylvania Press.

Nuryanti, W. (1996) 'Heritage and Postmodern Tourism', *Annals of Tourism Research* 23: 249–60.

Ochs, E. and Capps, L. (1996) 'Narrating the Self', *Annual Review of Anthropology* 25: 19–43.

Odermatt, P. (1996) 'A Case of Neglect? The Politics of (Re)presentation: a Sardinia Case', in J. Boissevain (ed.), *Coping with Tourists*, Oxford: Berghahn.

Okely, J. and Callaway, H. (1992) *Anthropology and Autobiography*, London: Routledge.

Öngör, A. (1999) 'The Turkish Economy Today', in D. Shankland (ed.), *The Turkish Republic at 75 Years*, Huntingdon: Eothen Press.

Öniş, Z. (1996) 'The State and Economic Development in Contemporary Turkey: Elitism to Neoliberalism', in V. Mastny and R.C. Nation (eds), *Turkey Between East and West: New Challenges for a Rising Regional Power*, Colorado: Westview Press.

Open Road Publishing Turkey Guide (1996) New York: Open Road Publishing.

O'Rourke, D. (1987) *Cannibal Tours* (film), Los Angeles: Direct Cinema.

Ousby, I. (1990) *The Englishman's England: Taste, Travel and the Rise of Tourism*, Cambridge: Cambridge University Press.

Pacal, J. (1996) 'Nature's lesson to the globe: Cappadocia', *Turkish Daily News*, 11 June.

Patullo, P. (1996) *Last Resorts*, London: Cassell.

Pearce, D. (1989) *Tourist Development*, London: Longman.

Pearce, P., Moscardo, G. and Ross, G. (1996) *Tourism Community Relationships*, Oxford and New York: Pergamon.

Peristiany, J.G. (ed.) (1966) *Honour and Shame – The Values of Mediterranean Society*, London: Weidenfeld & Nicolson.

—— (1968) *Contributions to Mediterranean Sociology*, Paris: Mouton & Co.

Picard, M. (1990) ' "Cultural tourism" in Bali: Cultural Performances as Tourist Attraction', *Indonesia* 49: 37–74.

—— (1993) 'Cultural tourism in Bali', in M. Hitchcock *et al.* (eds), *Tourism in South-East Asia*, London: Routledge.

—— (1995) 'Cultural Heritage and Tourist Capital: Cultural Tourism in Bali', in M. Lanfant, J. Allcock and E. Bruner (eds), *International Tourism – Identity and Change*, London: Sage.

—— (1996) *Bali: Cultural Tourism and Touristic Culture*, Singapore: Archipelago Press.

Picard, M. and Wood, R. (eds) (1997) *Tourism, Ethnicity, and the State in Asian and Pacific Societies*, Honolulu: University of Hawai'i Press.

Pine, J. and Gilmore, J.H. (1999) *The Experience Economy: Work is Theatre and Every Business a Stage*, Boston: Harvard Business School Press.

Pinney, C. (1994) 'Future Travel', in L. Taylor (ed.), *Visualising Theory*, London: Routledge.

Pitt-Rivers, J. (1966) 'Honour and Social Status', in J.G. Peristiany (ed.), *Honour and Shame – The Values of Mediterranean Society*, London: Weidenfeld & Nicolson.

—— (1968) 'The Stranger, the Guest, and the Hostile Host: Introduction to the Study of the Laws of Hospitality', in J.G. Peristiany (ed.), *Contributions to Mediterranean Sociology*, Paris: Mouton & Co.

Plachter, H. and Rossler, M. (1995) 'Cultural Landscapes: Reconnecting Culture and Nature', in B. von Droste, H. Plachter and M. Rossler (eds), *Cultural Landscapes of Universal Value*, Jena: Gustav Fischer.

Poon, A. (1994) 'The "new tourism" revolution', *Tourism Management* 15: 91–2.

Pope, N. and Pope, H. (1997) *Turkey Unveiled – Ataturk and After*, London: John Murray.

Pretes, M. (1995) 'Postmodern Tourism', *Annals of Tourism Research* 22: 1–9.

Pruitt, D. and LaFont, S. (1995) 'For Love and Money – Romance Tourism in Jamaica', *Annals of Tourism Research* 22: 422–40.

Puijk, R. (1996) 'Dealing with Fish and Tourists: A Case Study from Northern Norway', in J. Boissevain (ed.), *Coping with Tourists*, Oxford: Berghahn Books.

Rabinow, P. (ed.) (1984) *The Foucault Reader*, London: Routledge.

Radcliffe-Brown, A.R. (1940) 'On Joking Relationships', *Africa* 13(3): 195–210.

Redfield, R. (1960) *The Little Community and Peasant Society and Culture*, Chicago: University of Chicago Press.

Redfoot, D. (1984) 'Tourist Authenticity, Touristic Angst, and Modern Reality', *Qualitative Sociology* 7(4): 291–309.

Reed, M. (1997) 'Power Relations and Community-Based Tourism Planning', *Annals of Tourism Research* 24: 566–91.

Richards, G. (1996) 'Production and Consumption of European Cultural Tourism', *Annals of Tourism Research* 23: 261–83.

Riggins, S. (ed.) (1990) *Beyond Goffman – Studies on Communication, Institution and Social Interaction*, Berlin and New York: Mouton de Gruyter.

Riley, P.J. (1988) 'Road Culture of International Long-term Budget Travelers', *Annals of Tourism Research* 15: 313–28.

Ritzer, G. and Liska, A. (1997) ' "McDisneyization" and "Post-Tourism": Complementary Perspectives on Contemporary Tourism', in C. Rojek and J. Urry (eds), *Touring Cultures – Transformations of Travel and Theory*, London: Routledge.

Rivers, P. (1973) 'Tourist Troubles', *New Society* 23: 539.

Robertson, G. *et al.* (eds) (1994) *Travellers' Tales – Narratives of Home and Displacement*, London: Routledge.

Robertson, R. (1992) *Globalisation – Social Theory and Global Culture*, London: Sage.

Robins, K. (1996) 'Interrupting Identities – Turkey/Europe', in S. Hall and P. du Gay (eds), *Questions of Cultural Identity*, London: Sage.

Robinson, H. (1976) *A Geography of Tourism*, London: Macdonald & Evans.

Rojek, C. (1995) *Decentring Leisure. Rethinking Leisure Theory*, London: Sage.

Rojek, C. and Urry, J. (eds) (1997) *Touring Cultures – Transformations of Travel and Theory*, London: Routledge.

Rosenwald, G.C. and Ochberg, R.L. (eds) (1992) *Storied Lives*, New Haven: Yale University Press.

Sack, R.D. (1992) *Place, Modernity, and the Consumer's World*, London: The Johns Hopkins University Press.

Sahlins, M. (1974) *Stone Age Economics*, London: Tavistock.

Said, E. (1978) *Orientalism*, London: Routledge.

Sant Cassia, P. (1999) 'Tradition, tourism and memory in Malta', *The Journal of the Royal Anthropological Institute* 5: 247–63.

Scheffelin, E. (1998) 'Problematizing performance', in F. Hughes-Freeland (ed.), *Ritual, Performance, Media*, London and New York: Routledge.

Schiffauer, W. (1993) 'Peasants Without Pride', in P. Stirling (ed.), *Culture and Economy – Changes in Turkish Villages*, Huntingdon: Eothen Press.

Schiffer, R. (1982) *Turkey Romanticised – Images of the Turks in Early 19th Century English Travel Literature*, Bochum: Studienverlag Brockmeyer.

Schneider, M. (1993) *Culture and Enchantment*, Chicago: University of Chicago Press.

Schoss, J. (1995) 'Beach Tours and Safari Visions: Relations of Production and the Production of Culture in Malindi, Kenya', unpublished thesis, University of Chicago.

Scott, J. (1995) 'Sexual and National Boundaries in Tourism', *Annals of Tourism Research* 22: 385–403.

—— (1997) 'Chances and Choices: Women and Tourism in Northern Cyprus', in T. Sinclair (ed.), *Gender, Work and Tourism*, London: Routledge.

Seaton, A.V. *et al.* (eds) (1994) *Tourism – The State of the Art*, Chichester: Wiley.

Seidler, V. (1987) 'Reason, Desire and Male Sexuality', in P. Caplan (ed.), *The Cultural Construction of Sexuality*, London: Tavistock.

Selwyn, T. (1993) 'Peter Pan in South-East Asia: the brochures', in M. Hitchcock *et al.* (eds), *Tourism in South-East Asia*, London: Routledge.

—— (1994) 'The Anthropology of Tourism – The State of the Art', in A.V. Seaton *et al.* (eds), *Tourism – The State of the Art*, Chichester: Wiley.

—— (1996) 'Atmospheric Notes from the Fields: Reflections on Myth-collecting Tours', in T. Selwyn (ed.), *The Tourist Image: Myths and Myth-Making in Tourism*, New York and London: John Wiley & Sons.

—— (2001) 'Tourism Development, and Society in the Insular Mediterranean', in D. Ioannides, Y. Apostolopoulos and S. Sonmez (eds), *Mediterranean Islands and Sustainable Tourism Development: Practices, Management and Policies*, London: Continuum.

Seremetakis, C.N. (1994) 'The Memory of the Senses', in L. Taylor (ed.), *Visualising Theory*, London: Routledge.

Sezer, H. and Harrison, A. (1994) 'Tourism in Greece and Turkey: An Economic View for Planners', in A.V. Seaton *et al.* (eds), *Tourism – The State of the Art*, Chichester: Wiley.

Shankland, D. (1993) 'Alevi and Sunni in Rural Anatolia', in P. Stirling (ed.), *Culture and Economy – Changes in Turkish Villages*, Huntingdon: Eothen Press.

—— (1999) 'Development and the Rural Community – Inspired Restraint', in D. Shankland (ed.), *The Turkish Republic at 75 Years*, Huntingdon: Eothen Press.

Sharpley, R. and Sharpley, J. (1997) 'Sustainability and the Consumption of Tourism', in M.J. Stabler (ed.), *Tourism and Sustainability*, Wallingford: CAB International.

Shaw, G. and Williams, A. (1994) *Critical Issues in Tourism – A Geographical Perspective*, Oxford: Basil Blackwell.

Shenhav-Keller, S. (1995) 'The Jewish Pilgrim and the Purchase of a Souvenir in Israel', in M. Lanfant, J. Allcock and E. Bruner (eds), *International Tourism – Identity and Change*, London: Sage.

Shilling, C. (1993) *The Body and Social Theory*, London: Sage.

Shooter, J. and Gergen, K.J. (eds) (1989) *Texts of Identity*, London: Sage.

Silver, I. (1993) 'Marketing Authenticity in Third World Countries', *Annals of Tourism Research* 20(2): 302–18.

Simmel, G. (1955) *Conflict and the Web of Group Affiliations*, London: Collier-Macmillan.

Sinclair, T. (1997) 'Issues and theories of gender and work in tourism', in T. Sinclair (ed.), *Gender, Work and Tourism*, London: Routledge.

Sirman, N. (1990) 'State, Village and Gender in Western Turkey', in A. Finkel and N. Sirman (eds), *Turkish State and Turkish Society*, London: Routledge.

Smith, V. (ed.) (1989) *Hosts and Guests*, 2nd edn, Philadelphia: University of Pennsylvania Press.

Smith, V. and Eadington, R. (eds) (1992) *Tourism Alternatives – potentials and problems in the development of tourism*, Philadelphia: University of Pennsylvania Press.

So, A.Y. (1990) *Social Change and Development – Modernisation, Dependency, and World-System Theories*, California and London: Sage.

Sontag, S. (1979) *On Photography*, Harmondsworth: Penguin.

—— (1983) *A Susan Sontag Reader*, New York: Vintage.

Spencer, C. (1993) *Turkey Between Europe and Asia*, Wilton Park Paper 72, London: HMSO.

Spurling, L. (1977) *Phenomenology and the Social World – The Philosophy of Merleau-Ponty*, London: Routledge & Kegan Paul.

Stabler, M.J. (ed.) (1997) *Tourism and Sustainability*, Wallingford: CAB International.

Starr, J. (1984) 'The Legal and Social Transformation of Rural Women in Aegean Turkey', in R. Hirschon (ed.), *Women and Property – Women as Property*, London: Croom Helm.

Steiner, C.B. (1994) *African Art in Transit*, Cambridge: Cambridge University Press.

Stendhal (1962) *Memoirs of a Tourist*, Evanston: Northwestern University Press.

Stiles, A. (1991) *The Ottoman Empire: 1450–1700*, London: Hodder & Stoughton.

Stirling, P. (1965) *Turkish Village*, London: Weidenfeld & Nicolson.

—— (1974) 'Cause, Knowledge and Change: Turkish Village Revisited', in L. Davis (ed.), *Choice and Change: Essays in honour of Lucy Mair*, London: The Athlone Press.

—— (1993) 'Intoduction', in P. Stirling (ed.), *Culture and Economy – Changes in Turkish Villages*, Huntingdon: Eothen Press.

Stokes, M. (1992) *The Arabesk Debate – Music and Musicians in Modern Turkey*, Oxford: Clarendon Press.

Stoller, P. (1989) *The Taste of Ethnographic Things*, Philadelphia: University of Pennsylvania Press.

Stone, L. (1998) *Representations of Turkey*, Ankara: ATS.

Stonich, S. (1989) 'Political Ecology of Tourism', *Annals of Tourism Research* 25: 25–54.

—— (2000) *The Other Side of Paradise: Tourism, Conservation, and Development in the Bay Islands*, New York, Sydney and Tokyo: Cognizant Communication Corporation.

Strang, V. (1997) *Uncommon Ground: Cultural Landscapes and Environmental Values*, Oxford and New York: Berg.

Swain, M. (1993) 'Women Producers of Ethnic Arts', *Annals of Tourism Research* 20: 32–52.

—— (1995) 'Gender in Tourism', *Annals of Tourism Research* 22: 247–66.

Synnott, A. (1993) *The Body Social*, London: Routledge.

Tapestry Holidays (1997) *Uncommercial Turkey 1997–1998* (brochure).

Tapper, N. (1991) *Bartered Brides – Politics, Gender and Marriage in an Afghan Tribal Society*, Cambridge: Cambridge University Press.

Taylor, L. (ed.) (1994) *Visualising Theory*, London: Routledge.

Tekeli, S. (ed.) (1995) *Women in Modern Turkish Society – A Reader*, London and New Jersey: Zed Books.

Teodor, S. (1971) *Peasants and Peasant Societies: Selected readings*, Harmondsworth: Penguin.

Tomlinson, J. (1991) *Cultural Imperialism*, Baltimore: The Johns Hopkins University Press.

Tosun, C. (1997) 'Questions about tourism development within planning paradigms: the case of Turkey', *Tourism Management* 18: 327–9.

—— (1998) 'Roots of unsustainable tourism at the local level: the case of Ürgüp in Turkey', *Tourism Management* 19: 595–610.

Towner, J. (1996) *An Historical Geography of Recreation and Tourism in the Western World, 1540–1940*, Chichester: Wiley.

Travel and Tourism Intelligence (2001) *TTI Country Reports*, Travel and Tourism Intelligence.

Trawick, M. (1990) 'The Ideology of Love in a Tamil Family', in O.M. Lynch (ed.), *Divine Passions: The Social Construction of Emotion in India*, Delhi etc.: Oxford University Press.

Tucker, H. (1997) 'The Ideal Village: interactions through tourism in Central Anatolia', in S. Abram, D. Macleod and J. Waldren (eds), *Tourists and Tourism – identifying with people and places*, Oxford: Berg.

—— (2000) 'Tourism and the Loss of Memory in Zelve, Cappadocia', *Journal of the Oral History Society* 28(2): 79–88.

—— (2001) 'Tourists and Troglodytes: Negotiating Tradition for Sustainable Cultural Tourism', *Annals of Tourism Research* 28(4): 868–91.

—— (2002) 'Welcome to Flintstones-Land: Contesting Place and Identity in Göreme, Central Turkey', in S. Coleman and M. Crang (eds), *Tourism: Between Place and Performance*, Oxford: Berghahn Books.

turizm.net (2002) 'Cappadocia: Göreme', *http://www.turizm.net/cities/cappadocia/goreme.htm* (accessed 2 March 2002).

Turkish Daily News (1997) 'A–Z of Cappadocia' (supplement to the *Turkish Daily News* in Nevşehir).

Turkish Ministry of Tourism (1995) *Ankara and the Central Anatolian Region* (promotional brochure, printed by Mega Print).

Turner, B. (1984) *The Body and Society*, Oxford: Basil Blackwell.

Turner, K. (1999) 'From Classical to Imperial: Changing Visions of Turkey in the Eighteenth Century', in S. Clark (ed.), *Travel Writing and Empire: Postcolonial Theory in Transit*, London: Zed Books.

Turner, L. and Ash, J. (1975) *The Golden Hordes*, London: Constable.

Turner, V. (1969) *The Ritual Process*, Chicago: Aldine.

—— (1979) *Process, Performance and Pilgrimage*, New Delhi: Concept.

Turner, V. and Bruner, E. (eds) (1986) *The Anthropology of Experience*, Chicago: University of Illinois Press.

Turner, V. and Turner, E. (1978) *Image and Pilgrimage in Christian Society – Anthropological Perspectives*, New York: Columbia University Press.

Udesky, L. (1997) 'Fieldwork and fantasy in Göreme Village', *Turkish Daily News*, 9 May.

Urry, J. (1988) 'Cultural Change and Contemporary Holiday-Making', *Theory, Culture and Society* 5: 35–56.

—— (1990) *The Tourist Gaze*, London: Sage.

—— (1992) 'The Tourist Gaze and the Environment', *Theory, Culture and Society* 9: 1–26.

—— (1995) *Consuming Places*, London: Routledge.

Van den Berghe, P. (1992) 'Tourism and the Ethnic Division of Labor', *Annals of Tourism Research* 19: 234–49.

—— (1994) *The Quest of the Other – Ethnic Tourism in San Cristobal, Mexico*, Washington, DC: University of Washington Press.

Van Gennep, A. (1960) *The Rites of Passage*, London: Routledge & Kegan Paul; 1st pub., 1909.

Van Nieuwenhuijze, C.O. (1997) 'Near Eastern Village: A Profile', in O. Ari (ed.), *Readings in Rural Sociology*, Istanbul: Bogaziçi University.

Veal, A.J. (1992) *Research Methods for Leisure and Tourism – A Practical Guide*, London: Pitman Publishing.

Veijola, S. and Jokinen, E. (1994) 'The Body in Tourism', *Theory, Culture and Society* 11: 125–51.

Wahnschafft, R. (1982) 'Formal and Informal Tourism Sectors – A Case Study in Pattaya, Thailand', *Annals of Tourism Research* 9: 429–51.

Waldren, J. (1996) *Insiders and Outsiders – Paradise and Reality in Mallorca*, Oxford: Berghahn Books.

—— (1997) 'We Are Not Tourists – We Live Here', in S. Abram, D. Macleod and J. Waldren (eds), *Tourists and Tourism – identifying with people and places*, Oxford: Berg.

Wallerstein, I. (1976) *The Modern World System: Capitalist Agriculture and the Origins of the European World Economy in the Sixteenth Century*, New York: Academic Press.

Wang, N. (2000) *Tourism and Modernity: A Sociological Analysis*, Oxford: Pergamon/Elsevier Science.

Wanhill, S. (2000) 'Small and Medium Tourism Enterprises', *Annals of Tourism Research* 27: 132–47.

WCED (1987) *Our Common Future*, Oxford: Oxford University Press (for World Commission on Environment and Development).

wec-net.com (2002) 'Saksagan Cave Hotel', *http://holiday.wec-net.com.tr/turtle* (accessed 2 March 2002).

Weightman, B.A. (1987) 'Third World Tour Landscapes', *Annals of Tourism Research* 14: 227–39.

Wheatcroft, A. (1995) *The Ottomans*, London: Penguin Books.

Wheeller, B. (1997) 'Here We Go, Here We Go, Here We Go', in M.J. Stabler (ed.), *Tourism and Sustainability*, Wallingford: CAB International.

Whelan, T. (ed.) (1991) *Nature Tourism. Managing for the Environment*, Washington, DC: Island Press.

Whiting, J. (1939) 'Where Early Christians Lived in Cones of Rock', *National Geographic*: 763–802.

Wickers, D. (1997) 'The Perfect Turkey', *The Sunday Times*, 2 February.

Williams, W. and Papamichael, E.M. (1995) 'Tourism and Tradition: Local Control versus Outside Interests in Greece', in M. Lanfant, J. Allcock and E. Bruner (eds), *International Tourism – Identity and Change*, London: Sage.

Wilson, D. (1997) 'Paradoxes of Tourism in Goa', *Annals of Tourism Research* 24: 52–75.

Wood, K. and House, S. (1993) *The Good Tourist Guide to Turkey*, London: Mandarin.

Wood, R. (1994) 'Some Theoretical Perspectives on Hospitality', in A.V. Seaton *et al.* (eds), *Tourism – The State of the Art*, Chichester: Wiley.

—— (1997) 'Tourism and the State: Ethnic Options and Constructions of Otherness', in M. Picard and R. Wood (eds), *Tourism, Ethnicity, and the State in Asian and Pacific Societies*, Honolulu: University of Hawai'i Press.

Woolcock, M. (1998) 'Social capital and economic development: Towards a theoretical synthesis and policy framework', *Theory and Society* 27: 151–208.

Woolgar, S. (ed.) (1988) *Knowledge and Reflexivity*, London: Sage.

WTO Seminar on Tourism Development in the Republic of Central Asia and the Black Sea (1993) Istanbul.

Yale, P. (1996) 'A Future for Göreme?', *In Focus Magazine* 20, London: Tourism Concern.

Zarkia, C. (1996) 'Philoxenia Receiving Tourists – but not Guests – on a Greek Island', in J. Boissevain (ed.), *Coping with Tourists*, Oxford: Berghahn Books.

Zeppel, H. (1998) 'Land and culture: sustainable tourism and indigenous peoples', in C.M. Hall and A. Lew (eds), *Sustainable Tourism: A Geographical Perspective*, Harlow: Longman.

Zinovieff, S. (1991) 'Hunters and Hunted: Kamaki and the Ambiguities of Sexual Predation in a Greek Town', in P. Loizos and E. Papataxiarchis (eds), *Contested Identities*, Princeton: Princeton University Press.

Index

Note: Page numbers followed by 'n' refer to notes

Accommodation Association (Office) 9, 47, 102, 112, 114
aesthetic 2, 3, 16, 28, 34–5, 160, 163–5, 178, 185; aestheticization 1, 2, 28, 163–4
Africa 45
Ankara 5, 14, 87, 89, 143, 164
anti-tourist 51–2, 54
Atatürk, Mustafa Kemal 11, 42n
Australia 155
Australian 33, 44, 45, 49–50, 109–10, 130–2, 144–5
authentically social 62–6, 120–2, 134, 172–3, 177–8, 182, 185
authenticity 2–3, 20, 21, 26–8, 34, 119–20, 123, 174–7, 183–6
Avanos 6, 10, 12–13, 158n

backgammon 48, 54, 109
backpackers 10, 44–51, 54, 97, 125, 131, 134–5, 178, 187
Bali 13, 116n, 173
ballooning: hot-air 72, 100, 108–9
bargaining 49, 59–60, 105, 112–14, 129–30, 136n
Baudrillard, J. 68n
Belgium 5, 72, 87
Bellér-Hann, I. and Hann, C. 11, 15, 90n, 92
Boissevain, J. 183
Boorstin, D. 68n
Bourdieu, P. 51, 90n, 157n
Bowman, G. 137–8, 148
British 44, 45, 47
brochure 23–5, 28, 30, 38, 42
Bruner, E. 3, 19, 22n, 37–8, 56, 64, 135n, 136n, 173
Bruntland Report 22n
building development 4, 6, 13, 160,

165–8; restriction 4, 6, 13, 164, 168; regulation 4, 29, 111, 161, 165, 180n, 181n
business competition 13–14, 16, 19, 21, 46, 91, 97, 98, 99, 103–16, 117n, 179
Byzantine 9, 24, 25, 26, 30–2, 35, 38, 42, 160, 168

Çannakale 44, 49
Cappadocia Preservations Office/Committee 8, 22n, 29, 111, 159–66, 168, 180n
carpet shop 9, 39, 47, 94, 100, 102
cave dwelling 2, 3, 5, 6, 24–5, 34–5, 38, 46, 48, 119–22, 160, 163, 168, 170–2, 176, 178
Central Anatolia 25, 27, 31, 38, 85, 140
Chambers, E. 2, 22n, 135n, 176, 185–6
change 1, 3, 91, 98, 115, 132, 157, 166–9, 173, 183, 185–6; cultural change 4, 22n, 132, 142, 186; social change 11, 15, 16, 20, 69, 74, 84–7, 92, 114, 149, 150, 155–7, 169, 183, 186, 188
Christian 24, 29–33, 36–8, 40, 42n, 168; church 26, 29–33, 35, 39, 41, 45, 46, 160, 187
climate 5, 12
coast: Turkey's south-west 36, 44, 47, 49
Cohen, E. 22n, 68n, 174, 181n
commission 19, 38–9, 97, 100, 102–3, 109, 112, 116, 117n
community-based tourism 4, 14, 92
consumption 34, 36
Co-operative, Göreme Tourism 100
Crick, M. 17, 116n, 117n, 135n
crop 5
cultural capital 52
cultural difference 3, 35, 42, 63, 163–4, 173, 177, 186

cultural form 1, 21, 183, 185–6
cultural homogenisation 3
cultural tourism 2, 11, 12, 26, 34, 44, 62, 118, 134, 177, 185–7
culture 2, 22n, 25, 35, 46, 47, 62, 118, 122, 163, 177, 179, 185

Dann, G. 51, 67n, 181n
Delaney, C. 22n, 78, 82, 90n, 116n, 158n
development: tourism 39, 167, 169, 182
Disney 171, 174–5; Disneyfication 178; Disneyland 169
domestic (Turkish) tourism 44, 99

Eco, U. 169, 174–5
economics 5, 82, 92, 129, 151–2, 179; agricultural economy 5, 11, 21, 69, 83–4, 91; economic development 11–12, 14, 85–7, 91, 179; formal/informal economy 99, 116n, 135n; liberalisation 11, 92
Elsrud, T. 51, 54, 67n
employment 101–3, 109–11, 144
entrepreneur 69, 91, 93–4, 99, 102, 106, 112, 115, 135n, 182
ethnography 16, 22n, 183; ethnographic approach 15; ethnographic knowledge 15; ethnographic method 17; ethnographic study 16, 17; ethnographic text 16
Europe 12, 44, 45, 48, 88, 155, 183
European Union 12
evil-eye 105–6, 117n

fairy-chimneys 5, 6, 23, 25–6, 28, 35, 38, 59, 75, 140, 159, 160, 161, 168–71, 175
farming (gardening) 5, 35, 69, 83–4, 101, 178
Featherstone, M. 175
Feifer, M. 125, 184, 186
festival 18, 126, 179–80
fieldwork 4, 15–20, 22n, 36, 73; gender issues in fieldwork 17–19, 77–8, 117n
Flintstones 1, 3, 49, 159, 169–80, 182, 186, 187
Foucault, M. 136n
France 44

Gallipoli 44
gender relations 15, 17, 21, 22n, 76–9, 137, 141–3, 145–52, 155–7, 158n,

179; gender segregation 8–9, 69, 74, 79–82, 141
Germany 5, 44, 46
gezmek 17, 77–82, 121, 141, 157
globalisation 2, 3, 161–2, 171, 186
Goffman, E. 117n
Göreme National Park 2, 4, 10, 13, 19, 25, 28–9, 34–5, 36, 39, 40, 92, 160, 161, 164, 187
Göreme Open-Air Museum 4, 9–10, 13, 31, 32, 36, 39, 41–3, 45–8, 92–3, 95, 100, 160, 168–9
Graburn, N. 68n, 131, 136n, 181n
grapes 5, 69, 70, 83
Greece 42n, 44, 129, 147, 180n
guidebook 23, 24, 30, 39, 42, 45, 47, 49, 67, 106, 140, 162
guided tour 39, 186; tour guide 19, 38–9, 46, 119–20, 183, 185
Gulf War 97

happenstance 55, 66–7, 184–5
heritage 2, 11, 23, 25–6, 31, 38, 42, 161
Herzfeld, M. 118, 122, 129, 132, 135n, 180n
Holland (Netherlands) 5, 44
Hollinshead, K. 180n
Hong Kong 44
honour 69, 73, 77–9, 90n, 103, 108, 122, 142–3, 147, 182
hospitality 21, 38, 107, 118, 122–35, 135n, 178, 179, 182–3, 188
host–guest relationship 4, 12, 21, 33, 98–9, 116n, 118, 122–8, 134–5, 179, 182–3
human rights 12
hyper-reality 175; *hypo*-reality 172, 175–7

identity 2, 3, 4, 21, 51–5, 57, 75–6, 116, 118, 121–2, 125, 134, 139, 156, 157, 160, 170–2, 177–80, 183–6
impact 1, 3, 20, 22n, 53
independent tourists 10, 12, 20, 39–42, 44, 118, 134, 183, 185–8; numbers of visiting Göreme 10, 22n
India 13, 38, 116n
individualism 52–4, 57, 67, 92, 114; individuating experience/identity 53, 55, 57, 66–7, 128, 134, 135, 184, 188
infidel 141
interaction 4, 20, 21, 41, 43–4, 49, 58, 62, 63, 66–7, 117n, 118, 123, 125–8, 134, 137, 140, 160, 172–3, 177, 178,

182–6, 188; observation of interaction 16, 17, 18–19
Internet 35, 42, 100, 173, 185, 188; Internet café 187
irony 53–4, 133, 170, 172, 176–8, 180, 183, 186
Islam 11, 15, 31, 69, 78, 90n, 124, 141, 142
Istanbul 10, 14, 36, 38, 44, 46, 49, 140, 143, 187

Japanese 44, 48, 53

Kandiyoti, D. 22n, 68n, 90n
Kohn, T. 138, 157n, 158n
Kurds (Kurdish) 12, 44, 49, 106, 148, 189

landscape 2, 24–9, 36, 38–9, 40, 42, 45–8, 58, 75, 159, 160, 168, 174, 178, 179; cultural landscape 34–6, 40, 161; tourist landscape 36–42, 187
legislation 12–13, 14, 99, 110–11, 165, 180n
liminality 139, 143, 169, 178, 181n, 183, 186
local participation 4, 14; *see also* tourism business, local ownership of
London 42n, 47, 49, 50, 143

MacCannell, D. 22n, 24, 34, 42, 65, 67n, 68n, 120, 133, 159, 184
Mallorca 180n
marriage 73, 76, 137, 140, 142–3, 150, 156–7, 157n, 158n
mass tourism 3, 13, 52–3, 67, 120, 187; *see also* package tourism
Mediterranean region 10–11, 105, 118; Mediterranean society 77, 90n
Middle East 45, 77, 87, 90n, 105, 143, 144–5
migration 5, 21, 72, 75, 85, 87–9, 91, 121, 143, 150, 154–5, 157, 170, 179
Ministry of Culture 4, 8, 19, 22n, 42n, 111, 159, 160, 161, 164–5, 169
Ministry of Tourism 14, 19, 25, 28, 180n
Ministry of Tourism and Culture 13
modernity 33, 34, 61, 65, 120, 166
Mowforth, M. 52–3, 162
Municipality (office) 161, 166–7, 180n, 181n, 187
Munt, I. 51–3, 57, 67n, 162, 181n
Muslim 5, 32, 33, 36, 42n, 76

NATO 12
negotiation 4, 16, 28, 43–4, 66–7, 105, 116, 120, 122, 123, 134–5, 148, 156, 160, 172, 177, 179–80, 183–6, 188
networks: business partnerships and alliances 100, 111–14, 117n
Nevşehir 5, 8, 9, 10, 12–13, 14, 25, 29, 37, 164, 187
New Zealanders 44, 45, 47–8, 50, 144

orientalism 12, 42, 161, 162
Ottoman 6, 11, 44; Empire 5

package tourism 10, 13, 19, 20, 36–9, 44, 56, 58, 61, 67, 118–20, 134–5, 185, 188; numbers visiting Göreme 9, 42n; numbers visiting Turkey 12; observation of 19, 119; package tourists 36–9, 40–1, 52, 183, 188
pansiyon 9, 12, 33, 35, 41, 45, 54, 95, 98–9, 100–2, 125–8, 160, 166, 169, 171, 179, 187
parody 133
participant observation 16–19
peasant society 22n, 75, 96, 105–6, 108–9, 116n
photography 24, 26, 38, 73–4, 121, 124–5; photographic gaze 136n
Picard, M. 1, 4, 13, 22n, 116n, 173
politics 4, 11–12, 15, 19, 160, 180n
post-modernism 65, 171–6, 186
post-tourist 125, 184, 186
preservation 2, 3, 6, 8, 13, 30, 33, 35, 111, 134, 159, 172–3, 176, 177, 179, 180n, 182, 183, 185–6

reflexivity 16, 22n
representation 4, 12, 14, 16, 19, 21, 23–4, 27–36, 39–40, 42, 43, 59, 102, 118, 119, 121, 134, 170, 172, 177, 182, 183, 186
restaurant 6, 9, 12, 14, 36, 49, 70, 93, 100
Riley, P. J. 51, 59, 67n, 136n
ritual 68n, 136n, 181n
romance 21, 137–40, 145–57, 157n
Romanticism 26, 33–5, 42, 64, 161, 162, 180n

Said, E. 161
Scott, J. 101
Selwyn, T. 1, 3, 22n, 23, 26, 62, 64, 178, 181n, 183, 185

serendipity 20, 43, 60, 66–7, 122, 128, 129, 135, 182–6, 189
sex tourism 157n
sexual relationship 21, 137–40, 146–8, 182
Shankland, D. 11
Singapore 44
souvenir 19, 38, 119, 129
Stirling, P. 8, 15, 22n, 75, 85, 90n, 96, 117n
subsistence 5, 69
sustainability 2, 3, 14, 19, 21, 22n, 92, 160, 177, 183, 185–8; sustainable tourism development 16, 162, 185

Thailand 13, 112
theme-park 1, 175, 182
tour agency 6, 9, 17, 36, 38–9, 41, 46, 47, 50, 56, 70, 94, 99, 183, 187; Göreme Tour Agency Association 112, 114
tourism business 4, 6, 8, 9, 16, 17, 87–8, 92–116, 137, 144, 152–6; local ownership of 4, 13–14, 21, 41, 101, 118, 126, 127, 135, 165, 179, 187, 188
Tourism Encouragement Act 12–13, 92
tourism industry 2, 23, 42, 67, 92, 186, 189
tourist accommodation 6, 9, 13–14, 35–6, 40–1, 49, 99, 163
tourist experience 3, 4, 16, 38, 39, 40, 42, 43–4, 55, 57–8, 62, 63, 65, 67, 120, 177–8, 183–6, 188
tourist gaze 4, 20, 21, 23, 28, 35–42, 43–4, 58–9, 66–7, 118, 121, 125, 134, 176, 177, 183–7, 188
tourist motivation 3, 45, 50, 66
touristic surrender 37–8, 56
touristification 1, 173, 177
traditional 2, 3, 35, 46, 63, 83, 102, 119,
120–2, 124, 141, 143, 172–3, 176, 179, 183, 185
transport 10, 22n, 44, 45, 47, 187–8
traveller–tourist distinction 51–62, 174, 177
troglodyte 5, 33, 35, 38, 40, 41, 42, 160, 170–2, 180, 183, 185, 186
Turkish government 11, 14, 35, 97, 160, 180n
Turkish Republic 5, 11, 32, 42n
Turner, V. 136n

underground city 36, 38, 46, 47, 49, 56
UNESCO (United Nations Educational, Scientific and Cultural Organization) 4, 31, 160, 161, 169
urbanisation 5, 11; urban society 158n
Ürgüp 4, 10, 12–14, 37, 44, 48–9, 53, 92, 97, 164
Urry, J. 4, 23, 28, 42n, 57, 65, 68n, 125, 136n, 180n, 181n, 184
USA 12, 35, 44, 48

Van Gennep, A. 68n, 136n, 181n
vernacular 6, 41, 186, 188
violence 109, 117n, 132, 146
virtual tourism 58, 178

Waldren, J. 180n
Wang, N. 4, 22n, 26–8, 34, 42n, 61, 68n
wedding 18, 73, 141, 156, 167, 180n, 182
Weightman, B. A. 38
women's relation to tourism 9, 15, 21, 69, 79, 82, 97, 101–2, 117n, 121, 149–56
World Heritage Site 2, 4, 31

Zelve (Open-Air Museum) 36, 42n, 43, 47, 48